ROADSIDE GEOLOGY OF OREGON

roadside
geology
of oregon

David D. Alt
Donald W. Hyndman

MOUNTAIN PRESS PUBLISHING CO.
Missoula, Montana

Sixthteenth Printing, July 2001

Library of Congress Cataloging-in-Publication Data

Alt, David D.
  Roadside geology of Oregon

  1. Geology—Oregon—Guide-books.   I. Hyndman, Donald W.,
joint author.   II. Title
QE155.A47   557.95   77-25841
ISBN 0-87842-063-0

Mountain Press Publishing Company
P.O. Box 2399
Missoula, MT 59806
(406) 728-1900

preface

We wrote this book for people who would like to know something about their geologic surroundings but don't have the time to wade through the scattered and often boring technical literature in which most of the information is securely hidden. We didn't intend this book for our professional colleagues who should know what they are looking at without our help but for people who don't have any background in geology and would like to know what they are looking at anyway. But we hope that a few of our colleagues will enjoy the book even though it wasn't intended for them.

Obviously we didn't figure all the geology in this book out all by ourselves. We read everything we could find about the geology of Oregon and then extracted the bits and pieces that seemed the most interesting, most important, and most visible. We tried to skim the cream off a century of geologic research in Oregon and we owe something to nearly everyone who has contributed to the great mass of Oregon geologic literature.

One of the problems in writing about geology as seen from the road is that half our readers will be going one way and the other half the other. There is no good way to write a description so it reads both ways at once. We tried to keep the people going the other way in mind and make things as easy for them as we could, but all our descriptions are written from north to south or from west to east. Presumably the people going the other way will come back some day.

Neither is it possible to write a description of every road in the state, or even every major road, and still wind up with one book instead of a shelf full of them. So we selected a network of roads that give a good representative view of the rocks. Many of our maps also include the neighboring roads and we hope our readers will learn to use just the maps without needing an accompanying description.

Geology is a matter of knowing what the rocks are and what they mean. We have tried to cover both aspects of the subject in this book, describing the rocks well enough that people can know them, or at least know a good many of them, and also understand how they formed and how they fit into the larger view of things.

The subject of geology has grown in its perception during the last 10 years, gaining its own larger view of many things for the first time. Many ideas now considered basic were still undreamed of only a decade ago. We have applied the new ideas in this book in order to make the discussion as modern as possible. Therefore, many of the broad interpretations in this book are our own and are published here for the first time.

David D. Alt and Donald W. Hyndman
Department of Geology
University of Montana
Missoula, Montana

contents

geologic

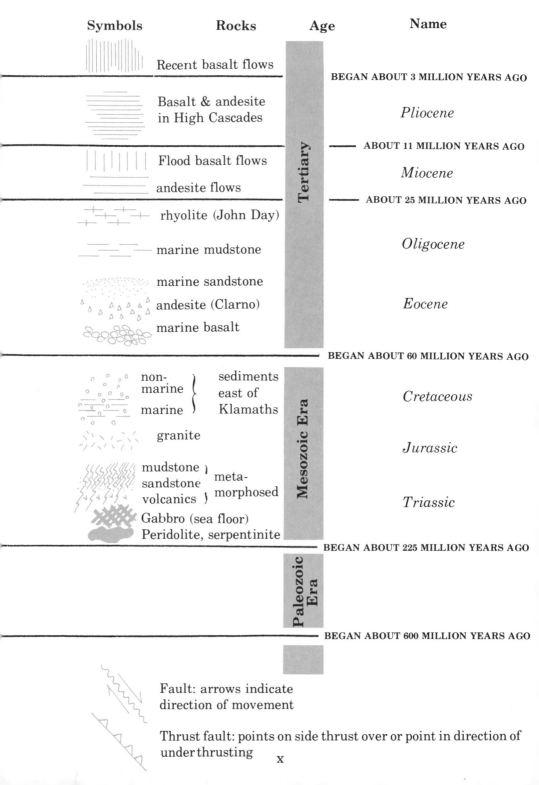

Symbols	Rocks	Age	Name

Recent basalt flows

BEGAN ABOUT 3 MILLION YEARS AGO

Basalt & andesite
in High Cascades

Pliocene

— **ABOUT 11 MILLION YEARS AGO**

Flood basalt flows

Miocene

andesite flows

— **ABOUT 25 MILLION YEARS AGO**

rhyolite (John Day)

marine mudstone

Oligocene

marine sandstone

andesite (Clarno)

Eocene

marine basalt

Tertiary

BEGAN ABOUT 60 MILLION YEARS AGO

non-marine ⎱ sediments
marine ⎰ east of Klamaths

Cretaceous

granite

Jurassic

mudstone ⎱
sandstone ⎰ meta-morphosed
volcanics ⎰

Triassic

Gabbro (sea floor)
Peridolite, serpentinite

Mesozoic Era

BEGAN ABOUT 225 MILLION YEARS AGO

Paleozoic Era

BEGAN ABOUT 600 MILLION YEARS AGO

Fault: arrows indicate
direction of movement

Thrust fault: points on side thrust over or point in direction of
under thrusting   x

time scale

Important events in Oregon

Eruption of High Cascade volcanoes, Ice ages, big lakes in southwestern Oregon

---

13-0? Block faulting, basalt and rhyolite eruption in southeastern Oregon; western part of Oregon pulled northward to give the state its present shape.

16-13 Eruption of plateau flood basalts
15 Gradual emergence of northern half of Coast Range

30-20 Volcanoes in Western Cascades, and light-colored ash of John Day Formation in Central Oregon
35 Seafloor sinking shifts from line curving northeast through Oregon to line offshore parallel to present coast.
50-40 Line of volcanoes to east also shifts.
Eruption of volcanoes in western Cascades and Clarno volcanoes in Central Oregon.
50 Coast Range near Coos Bay now emergent from sea.

---

70 Immense intrusion of granite in much of Idaho
100 Separation of Klamaths and Sierras into separate ranges.

150 Melting and intrusion of granite bathaliths in Klamaths

200 Descending sea floor scrapes sediments against continent to make Klamaths, Blocs, and Wallowas as coastal ranges. Limestone reefs fringe volcanic island arcs in Klamaths, Wallowas, and western Idaho.

---

300 "Continental" sandstones and marine limestone in southern Blue Mts.

400 Schists in eastern Klamaths of S. Oregon, N. California

 Basalt & andesite volcanoes

xi

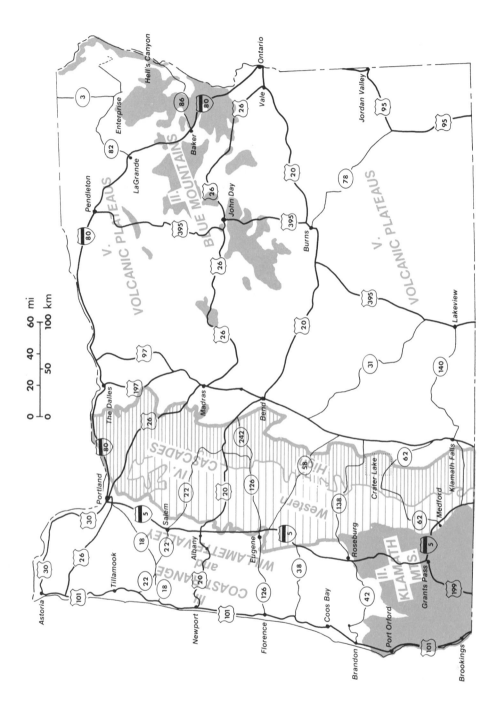

the plot

The geology of Oregon, indeed the very existence of Oregon, is the product of a prolonged collision between the North American continent and the Pacific Ocean floor which began about 200 million years ago and continues today. It is a long story with many involved details but a simple plot. We will begin by outlining the main plot in this first chapter and then deal with some of the details later in the chapters dealing with the various geologic regions of the state.

The earth has a rigid outer crust about 60 miles thick which consists of a patchwork mosaic of about a dozen or so pieces called "plates." These float on the denser rocks of the earth's interior which are so hot that they would melt were they not under the great pressure exerted by the weight of the overlying plates. But even though they are solid, the hot rocks of the earth's interior are soft and easily deformable; they flow under pressure as though they were modelling clay. So the crustal plates are not rigidly locked in position but move around on the soft rocks beneath them at varying rates which average an inch or two a year. We use the term "crust" a bit loosely here; geologists normally use the more cumbersome term "lithosphere."

Where moving crustal plates meet, they may slide past each other, pull away from one another, or simply collide. In places where two plates pull away from one another, the hot rock in the earth's interior partially melts to form basalt lava which wells up in the gap opening between them in much the same way that water in a lake will rise to fill the gap between two separating ice floes. As the molten basalt lava welling up between the separating plates and pouring out onto the nearby seafloor cools and solidifies, it becomes the upper part of the

1

rigid oceanic crust. The ocean floor is everywhere underlain by a thick sequence of basalt lava flows. Where two plates collide with one another, the heavier one sinks into the interior of the earth where it gets hot, softens and eventually loses its identity by the time it is about 200 miles down. So new crust is constantly forming where plates pull away from each other and old crust is constantly consumed into the interior of the earth where they collide.

Cross-section showing how crustal plates form, move away from each other, and finally sink someplace else.

Just a few hundred miles offshore from the coast of Oregon there is a rift where two crustal plates of the Pacific Ocean floor are pulling away from each other and new ocean floor is forming. The newly formed ocean floor is moving landward and sinking back into the earth's interior just off the Oregon coast. Meanwhile, the North American continent is moving westward, and probably somewhat southward as well, and the entire floor of the Pacific Ocean is moving northward. It is this set of circumstances that created Oregon — a glancing collision between the Pacific Ocean floor and the North American continent.

The stage was set for the action that created Oregon about 200 million years ago. Some parts of that earliest picture are clear enough but others are hazy, hidden under an overlay of more recent events and younger rocks. We can be quite sure that part of the western coast of North America was then approximately along the line of the present boundary between northeastern Oregon and Idaho and we can be sure that part of what is now the Klamaths existed then. It is possible that the Klamaths were an island standing offshore from a straight coastline but it seems far more likely that the coastline curved out toward them through what is now southeastern Oregon. If so, all the rest of Oregon was then a broad embayment of the ocean into the margin of the continent. Wherever the exact shorelines of that embayment may have been, we can be quite sure that they were fringed by broad coastal plains and continental shelves made of sands and muds eroded from the continent and then worked along the shore by the waves.

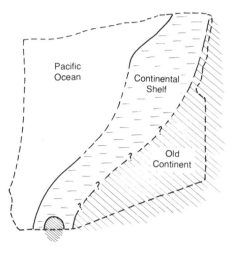

Map of Oregon as it probably existed about 200 million years ago. The outline of the state is distorted to correct for later movements.

Sometime around 200 million years ago the Pacific Ocean floor and the North American continent began to collide — they had been moving along together before then. It is hard to be sure what caused the collision but we do know that it began at about the time that North America separated from Europe and Africa and began to move westward as the Atlantic Ocean opened behind it. Perhaps that may have started the collision on the west coast. When the action began, the ocean floor began to sink beneath the westward moving continent along a line near the old coast.

When continents and ocean floors collide, it is always the ocean floor that sinks into the earth's interior because it is made of the heavier rocks. Continents are made of much lighter rocks so they always remain floating on the surface of the earth. The soft sands and muds that make the coastal plain and continental shelf and blanket the seafloor are also too light to sink into the earth's interior so the descending seafloor scrapes them off against the edge of the continent to make a coastal mountain range. We see remnants of that first coastal range today in the Blue and Wallowa Mountains of eastern Oregon and in most of what is now the Klamaths. The original Klamath island is only a small part of the present range. These mountains, like most coastal mountain ranges, are composed of an incredible mess of crushed sedimentary rocks originally deposited in widely separated areas of the coastal plain, continental shelf and seafloor and now all jammed together.

Coastal mountain ranges grow as long as the sinking seafloor continues to scrape its burden of muddy sediments off onto the edge of the continent. Slice after slice of mudstone and sandstone stuffed under the seaward edge of the range build it wider and higher as the constant movement slowly mangles all the rocks almost beyond recognition. Occasionally a generous slice of the bedrock seafloor itself manages to get incorporated into the growing coast range instead of sinking down into the interior of the earth as it is supposed to do. The final result is always a geologic nightmare, a chaotic mixture of a wide vari-

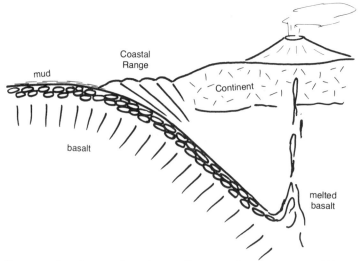

Cross-section showing how descending oceanic crust scrapes off its covering of mud to form a coastal range and then partially melts farther down to generate an inland chain of volcanoes.

4

ety of rocks originally formed at different times, in different ways, and at widely separated places all swept together into a hopelessly confused heap.

Wherever we see a coastal mountain range forming, we always find it paralleled about 100 miles inland by a chain of volcanoes. We see this arrangement today in the modern Coast and Cascade ranges and a similar situation existed when the Blue, Wallowa and Klamath Mountains were all part of the same coastal range. The basalt lava flows that cover the seafloor, the same lava flows that welled up in the rift between the separating plates and poured out onto the seafloor, melt when they get about 60 miles or so down into the earth and partly return to the surface through volcanoes.

The old volcanoes that once fringed the Klamaths, Wallowas and Blues are gone now, long since sacrificed to the slow processes of erosion. But we have certain evidence of their former existence in the great masses of granitic rocks incorporated in those ranges. Granitic rocks were once molten; they form when the extremely hot molten basalt rising from deep below the crust gets into the accumulating mass of muddy sediments which have a much lower melting point. The basalt melts large quantities of the sediments and may partly mix with them to form a melt that has the chemical composition of granitic rocks. As the molten magma rises toward the surface, some of it erupts through volcanoes to form the common gray or brown volcanic rock we call andesite and the rest slowly cools below the surface to form great masses of coarse-grained granitic rock. So the masses of granite we see today are the remains of magma reservoirs which once fueled big volcanoes.

So for a very long time, Oregon consisted of an accumulating coastal range separating a large embayment of the Pacific Ocean on the west from a parallel chain of volcanoes to the east and south. The bay was accumulating sediment washed into it by rivers draining the continent but the seafloor on which the sediment settled was moving landward, transporting it back into the growing coastal range. Where we see those sediments today, we find them full of volcanic debris — further evidence of those old volcanoes.

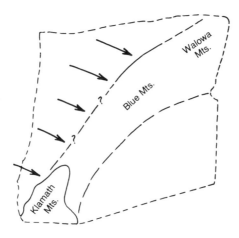

Map showing how Oregon probably looked about 150 million years ago when the Blue, Wallowa, and Klamath Mountains were actively growing coastal ranges. The arrow shows the probable direction of seafloor movement. Distortion of the state's outline is to correct for later crustal movements.

This same general situation persisted for a very long time starting between 200 and 150 million years ago, sometime during either the Triassic or Jurassic period, and lasting until about 30 million years ago, middle Oligocene time. The dinosaurs were just beginning to flourish when this phase of Oregon's development began and they developed to their maximum, disappeared from the earth, and had been extinct for nearly 40 million years before it ended. During those 150 or so million years, the coastal range grew slowly seaward and the line along which the seafloor was sinking into the interior also migrated slowly seaward, shifting the chain of volcanoes with it. As the volcanoes worked their way seaward, they loaded the older parts of the coastal range with great masses of granitic rocks. That great bay of ocean was slowly disappearing as the edge of the continent marched into it.

All that we see today of that old coastal range are the Blue, Wallowa, and Klamath Mountains; the rest is almost entirely buried beneath younger volcanic rocks. But the very nature of these younger volcanics betrays the existence and distribution of the older sedimentary rocks beneath them so we can be fairly confident that they exist even though they are buried. Evidently, the Blue, Wallowa, and Klamath Mountains are the only parts of that old coastal range high enough to stand above the floods of younger volcanic rock that eventually inundated most of Oregon. No one can be sure why these mountains grew so much higher than the rest but it seems reasonable to guess that they may simply mark the areas where the old seafloor had accumulated its greatest burdens of sediment before the big collision began.

It is possible, indeed quite likely, that some parts of that old coastal range never grew high enough to emerge above sea level and that the volcanoes may have marked the coastal rim of Oregon, perhaps as a curving row of volcanic islands. That seems to have been the case near the end of this prolonged first phase of Oregon's development about 40 million years ago. We find volcanic rocks of that age in the southern part of the present western Cascades about as far north as Eugene where they seem to swing northeastward. They are covered by younger rocks between Eugene and Redmond where they reappear in the Ochoco Mountains. We can follow them northeastward from the Ochocos in local patchy exposures and finally lose them completely in the area south of Pendleton.

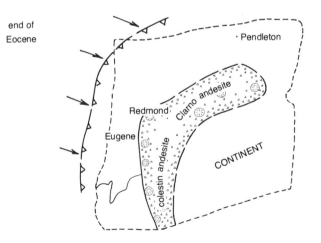

Map showing Oregon as it probably was about 40 million years ago when a chain of volcanoes and volcanic islands formed the outer rim of the state. Once again, the outline of the state is distorted to correct for later crustal movements.

The geologic setting changed abruptly about 35 million years ago when the seafloor stopped sinking along the curving line it had been following and began to sink along a line paralleling the present coast and some miles offshore. It is still sinking along that new line today.

The immediate effect of that sudden change was to halt the steady landward movement of the segment of seafloor between the old line of sinking and the new, creating a stable platform on which to build the rest of the state. Another effect was to shift the line of volcanoes from the curving course it had been following to the north-south line of the western Cascades.

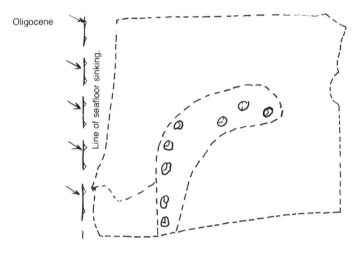

Map showing how the line of seafloor sinking shifted to essentially its present position about 35 million years ago. At this time all the old seafloor landward of the new line of sinking became stationary.

Within about 10 million years after the line of seafloor sinking shifted, the newly descending slab got deep enough into the earth's interior to begin melting. This started a new phase of volcanic activity along the line of the present western Cascades which quickly erected a row of volcanic islands across the mouth of the old ocean bay, converting what was left of it into an inland sea. Although we have no exposures of sediments deposited in that sea, we can be fairly sure that there must be a fairly thick accumulation of them because they would have come from both the continent to the east and the new volcanoes to the west.

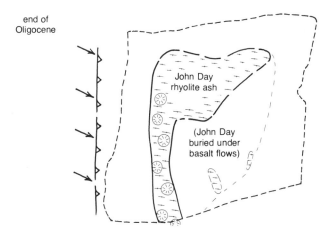

Map showing the new line of volcanoes built around 25 million years ago isolating much of the future state of Oregon as an inland sea. The outline of the state is still distorted.

Some of the activity in that first chain of western Cascade volcanoes was truly incredible. The volcanoes south of Eugene were built on rocks of the old coastal range, the Klamath Mountains, that cooked up into tremendous batches of granitic magma which erupted in explosive outbursts far exceeding any witnessed anywhere in the world during historic time. Searing clouds of steam and shreds of molten rock surged explosively all the way into central Oregon and other eruptions sent enormous clouds of light-colored airfall ash to blanket much of the region. That was an impressive episode of volcanic activity that must have lasted nearly 10 million years.

And then about 20 or 25 million years ago, in early Miocene time, the volcanic activity in the western Cascades abruptly stopped and a completely different kind of eruptions began, first in central and then in eastern Oregon. The new volcanoes in north-central and northeastern Oregon erupted nothing but basalt, incredible floods of it. Some individual eruptions produced hundreds of cubic miles of basalt in single lava flows that covered thousands of square miles. Again, nothing even remotely comparable has been witnessed in historic time. The new volcanoes in southeastern Oregon produced not only basalt but light-colored volcanic ash as well — evidence that the old coastal range does indeed underlie that part of the state.

The basalt floods covered almost all of Oregon east of the Cascades except for the higher parts of the Blue and Wallowa Mountains. They even filled what was left of the old inland sea and for the first time most of what is now Oregon was dry land, actually a volcanic plateau. Those lava floods also covered most of the older rocks in the state making it very difficult now to decipher the earlier stages in Oregon's development.

Then the flood basalt eruptions stopped and volcanic activity resumed in the western Cascades at about the end of Miocene time, about 12 million years ago. So there was a period when the volcanic activity shifted from the Cascades to the central and eastern part of the state and then shifted back to the Cascades again. This is a very strange sequence of events, something very peculiar must have happened in the interior of the earth beneath Oregon to cause it.

There is no way anyone can be sure what happened underneath Oregon to cause the unusual sequence of volcanic events during Miocene time. But it is interesting to speculate. One very important clue is the fact that large basalt eruptions seem always to occur in places where the upper crust of the earth cracks under tension. This relieves the pressure on the hot rocks in the earth's interior permitting them to partially melt to produce molten basalt. So any speculation on what may have happened must account for the halt in activity in the western Cascades and also explain what could have caused the crust in central and eastern Oregon to stretch and crack shortly afterwards. And then it has to go on to explain why the stretching stopped and activity resumed in the western Cascasdes.

We suggest that the descending slab of seafloor may simply have broken off and that the broken portion, being no longer attached to the seafloor still at the surface, may have sunk more rapidly into the earth's interior. If so, the soft rocks beneath the crust would have flowed rapidly, filling in behind the sinking slab, and much of that flowage would have had to come from beneath central and eastern Oregon. The resulting westward movement of material beneath the crust would have pulled Oregon westward, perhaps opening the cracks that produced the basalt floods. As soon as the broken slab had passed from beneath the western Cascades, activity there would stop and not resume until a new slab of seafloor had descended to that depth. So it is at least conceivable, although certainly not proveable, that a broken slab of seafloor beneath Oregon may have caused the strange sequence of Miocene volcanic events.

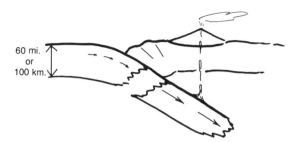

1) The descending slab breaks free and begins to sink more rapidly.

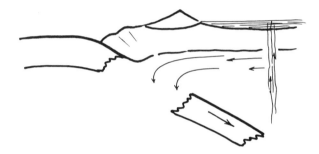

2) Activity ceases in the western Cascades because there is no longer a sinking slab of seafloor beneath them and material flows westward beneath Oregon to fill behind the sinking slab. Flood basalts erupt in central and eastern Oregon.

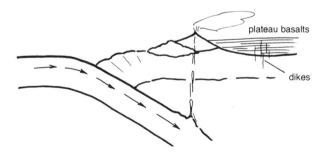

3) A new slab of seafloor descends to the level beneath the western Cascades to put them back in business; slowing westward movement of rock beneath Oregon halts the flood basalt eruptions.

The Cascades have been intermittently active since the end of Miocene time, ever since the flood basalt eruptions ceased; and there are a variety of other reasons for believing that a slab of seafloor is now descending beneath Oregon. The last million or so years has been a period of especially intense activity in the Cascades which has built the modern range of high volcanoes, many of which appear still active today. But none of the more recent eruptions even approaches the scale of those of 25 million or so years ago.

An interesting combination of crustal movement along faults accompanied by volcanic activity has been going on in south-central and southeastern Oregon for the last 10 to 15 million years and continues today. It is clear from the pattern of the fracturing and the direction the fault blocks are moving that the western part of the state has been moving northward relative to the eastern part — with the displacement distributed through hundreds of faults spread over an east-west distance of at least 200 miles. No one has figured out the total amount of movement; that would be a very tricky job. But it must be measurable in tens of miles, probably about 50 miles.

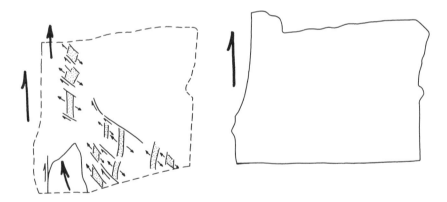

A pair of maps showing how fault movements have carried the western part of Oregon northward during the last 10 or 15 million years bringing the outline of the state to its present form.

Evidently the floor of the Pacific Ocean is moving northward relative to the North American plate, dragging the western edge of our continent along. In California this movement is

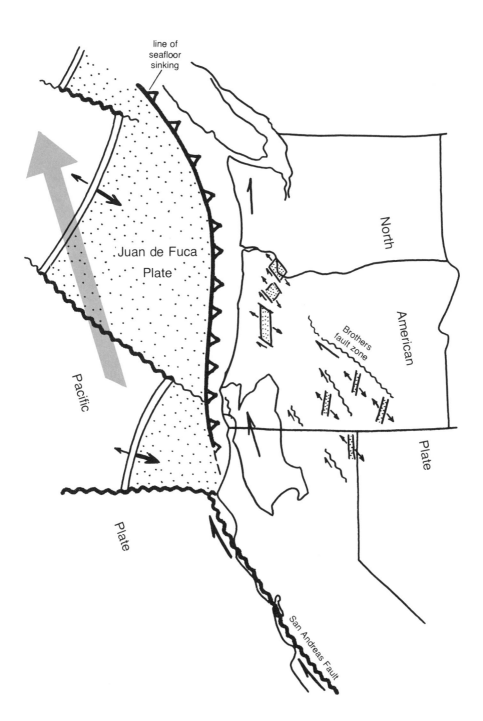

Map showing how the relative northward movement of the Pacific Ocean floor is differently expressed north and south of Cape Mendocino. Stippling shows the descending seafloor slab.

expressed in the notorious San Andreas fault which moves a western slice of the state northward at a rate of about 2 inches a year. But in Oregon and northernmost California the movement is not confined to a narrow fault zone but is instead distributed through hundreds of faults spread over a wide east-west distance. Perhaps this difference is caused by the cold slab of Pacific Ocean floor sinking beneath Oregon and northernmost California; perhaps the cold slab connects the two crustal plates across a broad zone thus distributing the movement. There is no cold slab now sinking beneath California south of Cape Mendocino so the connection between the two plates is only where they meet and the movement south of the cape is along a narrow instead of a wide zone of faulting.

Meanwhile, the Pacific Ocean floor continues to sink into the earth's interior along a line approximately 50 miles offshore from the present coast of Oregon. It continues to stuff its burden of muddy sediment under the edge of the big slab of ocean floor that stopped moving about 40 million years ago when the line of sinking shifted to its present location. As the sinking seafloor slides beneath the edge of that stationary slab, tucking more and more sediment under its western edge, the slab slowly rises while a new coastal range of wildly chaotic rocks forms offshore. The present Coast Range is essentially the uplifted western edge of that stationary slab of seafloor, not the same sort of mess we see in the Blue, Wallowa and Klamath Mountains. That mess is still forming beneath sea level some miles offshore and will eventually emerge to move the coast of Oregon another step seaward.

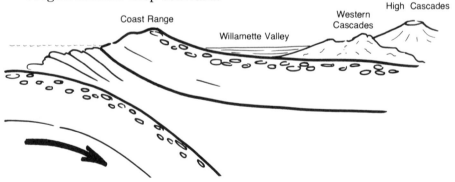

Section through the Coast Range and Cascades as they exist today.

The southern end of the Coast Range, the area around Coos Bay, just north of where the Coast Range meets the Klamaths, seems to have been above water since Eocene time, for the last 50 or so million years. The northern end, from around Newport to Astoria, seems to have been submerged as recently as Miocene time, about 15 or 20 million years ago. Evidently emergence of the Coast Range has been a slow process starting at the south and working gradually northward. This may well be due to the fact that the southern end is closer to the old land mass and therefore received a greater load of sediment than the northern end which was ocean floor fairly remote from land.

The emergence of the Coast Range on the western border of the state defined the Willamette Valley as a lowland between the Coast Range on the west and the Cascades to the east. It must have started as a bay which gradually filled with sediment and occasional lava flows until its floor was above sea level. Like the Coast Range, the floor of the Willamette Valley seems to have emerged above sea level at its southern end first and its northern end last.

The collision between the Pacific Ocean and North American plates that created Oregon still continues as actively as ever so the story isn't nearly finished. The seafloor sinking off the modern coast is stuffing its load of sediments under the edge of the older, now stationary, slab of seafloor that we see exposed in the present Coast Range. Eventually, when the offshore sinking stops, that mess now accumulating offshore will rise above sea level and the Oregon Coast Range will then resemble the one in central California where the sinking stopped long ago. The Cascades will continue to be intermittently active as long as the slab of seafloor continues to sink under Oregon and the western side of the state will continue to drag northward as long as the Pacific Ocean plate continues to move that way.

There are certain changes that we can anticipate if the present situation lasts another 15 or so million years. The rift where two parts of the Pacific Ocean floor are pulling away from each other is now very close to the coast of Oregon and Washington. If the North American plate continues to move westward as it is now doing, it will reach the position of that rift and the ocean-floor plate on the eastern side of it will disappear completely as the last of it sinks back into the earth's interior beneath the continent. When that happens, the rift will cease to exist and shortly thereafter there will be no descending slab of seafloor beneath Oregon. The Cascades will cease to erupt. The North American plate will then be attached to the Pacific Ocean plate on the western side of that rift, instead of on the eastern side as it now is. If the Pacific Ocean plate is still moving northward then, it will continue to drag the western part of Oregon along with it but the connection will then be just at the edges of the plates where they meet instead of through a broad zone of overlap where one is sinking beneath the other as is now the case. The effect will be to shift the northward movement to a narrow slice of westernmost Oregon instead of distributing it across the whole state. That is how the San Andreas fault got started in California — someday Oregon may have its own version of the San Andreas.

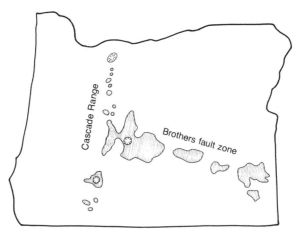

Distribution of modern volcanic activity in the Cascades and along the Brothers Fault zone.

the old mountains

A few of the earliest beginnings of Oregon are still visible but most are hidden now by later events so their reconstruction is partly a matter of conjecture and partly quite beyond recovering. Nevertheless, enough of the oldest rocks remain to give us at least a glimpse into the shadows of Oregon's most distant past. As geologic events are reckoned, those earliest beginnings are very recent indeed, hardly anything of what is now Oregon existed much more than about 200 million years ago.

We know that the edge of the North American continent was just east of the boundary between northeastern Oregon and Idaho some 200 million years ago because we can see the old rocks clearly exposed in the Salmon River canyon near Riggins, Idaho. And we know that the rocks in the core of the Klamaths, those in the very center of the Siskiyou Mountains along the boundary between California and Oregon, existed more than 200 million years ago. We also know that those old rocks in the Klamaths are now at least 60 miles west of where they used to be; but more of that later. The edge of the old continent is hidden between the Klamaths and northeastern Oregon but there are reasons for believing that it must have curved smoothly southwestward from northeastern Oregon through the original position of the Klamaths and then continued southward through the Sierra Nevada of California. The outline of that old continental edge is the beginning of the geologic story of Oregon.

The upper part of the earth's crust is quite different beneath continents and oceans. Continents are basically big slabs of light rocks, of granitic rocks complexly mixed with old sedimentary rocks mostly transformed into metamorphic rocks by prolonged cooking at high temperatures. Most continental slabs are about 25 miles thick and they float on the heavier rocks of the earth's interior for the same reason that wood floats on water — because they are light. The muddy sediments on the ocean floor rest on basalt lava flows which in turn rest on serpentinite, a green rock that forms when the heavy black rocks of the earth's interior absorb water. The heavy rocks of the earth's interior, they are called peridotite, are stiff and solid to a depth of about 60 miles, the base of the crust, where they pass downward into much weaker rocks which may even be partially melted.

Wherever exactly it may have been, the old continental margin that existed in Oregon before about 200 million years ago seems to have been stable for a very long time, probably for some hundreds of millions of years. Evidently the continent and seafloor were firmly attached to each other and riding along together with no relative motion between them. Where the boundary between a continent and seafloor is locked and stable, sediments eroded off the continent can accumulate to great depths to form extensive continental shelves and coastal plains. The Atlantic and Gulf coasts of the United States are in that condition today, essentially the same condition eastern Oregon must have been in for a very long time up until about 200 million years ago.

The firm connection between the continent and the ocean basin broke about 200 million years ago and the seafloor began to sink into the interior of the earth, sliding down beneath the edge of the continent. It is possible for the bedrock seafloor to do this because it is made of heavy rocks but continents are made of lighter stuff and always continue to float.

It is impossible to be very positive about exactly what may have happened some 200 million years ago to set the geologic story of Oregon in motion. We can be sure that the rocks beneath the earth's crust are moving about, rising in some places as they accumulate heat released by decay of radioactive elements and sinking in others as they lose heat to the surface. No doubt it was some change in this internal circulation within the earth that started things going in Oregon but there is no way of knowing what may have caused it.

It is interesting and probably significant to our story that North America separated from Europe and Africa and the Atlantic Ocean began to widen between them at about the same time events began in Oregon. Since then, our continent has moved about 1500 miles west of where it had been, half the width of the Atlantic Ocean, and it seems safe to assume at least that amount of Pacific Ocean floor has disappeared beneath its western edge to be consumed back into the interior of the earth. That seems unbelievable but these processes normally go at a rate of an inch or two each year and 200 million years is plenty of time.

The first thing the seafloor did as it began its long slide beneath the continent was to telescope together the sedimentary rocks of the old coastal plain and continental shelf, rudely jamming them against the edge of the continent to make the Blue, Wallowa, and Klamath Mountains. These ranges all consist of those old sediments now so scrambled together that there seems little hope that anyone will ever be able to sort them out except in a very general way.

If the margin of the old continent did indeed curve westward from northeastern Oregon toward the Klamaths, then a range of coastal mountains should have formed all along its edge. But all we see today are the Blue and Wallowa Mountains in northeastern Oregon and the Klamath Mountains in the southwestern corner of the state with none between. It is true that all of southeastern Oregon is covered by much younger volcanic rocks but if high mountains had formed in that area, they should rise above the volcanics as the Blue and Wallowa Mountains do. Evidently no high coastal mountains ever stretched across southeastern Oregon. But there might have been low ones lacking the stature to stand above the later floods of volcanic rocks. Coastal mountains don't necessarily have to be very high.

In fact, some of those younger volcanics may reveal almost as much as they conceal. Many of the volcanic rocks in southeastern Oregon are rhyolites which have a very high silica content and can form only by melting rocks already rich in silica. Old continental rocks and the crushed sediments in coastal mountains both contain enough silica to form rhyolite if they melt but the rocks that lie beneath the seafloor do not. So we can be sure that there are silica-rich rocks beneath southeastern Oregon and they must be old continent or younger coastal mountains. Since coastal mountains usually form along the edge of an older continent, it seems most likely that they include both.

Had the sedimentary accumulations been larger along some parts of the old continental margin than others, then larger coastal mountains would have formed in those places than in others. Perhaps the Blue and Wallowa Mountains and the Klamath Mountains simply mark areas where the old continental margin had developed an unusually large coastal plain and continental shelf, perhaps those large mountains are simply the crumpled remains of old river deltas or large coastal capes. Perhaps high mountains never formed along the reach of coastline that must have stretched through southeastern Oregon simply because the continental shelf and coastal plain were not very large in that area.

There is considerably more to formation of a coastal mountain range than simply jamming an old coastal plain and continental shelf into the continental margin. As mile after mile of seafloor continues to move slowly landward and slide beneath the continent, it continues to bring in more and more sediment from farther and farther offshore. The sediments can not possibly sink into the earth's interior with the seafloor they are riding because they are too light. So slice after slice of muddy sediment is scraped off the sinking seafloor and added to the seaward side of the growing coastal range; most of them are tucked under its seaward edge. Sometimes big slabs of the bedrock seafloor itself break loose and get shuffled into the growing mass of sediments in the coastal range instead of sliding peacefully beneath the continent as they are supposed to. The result is a chaotic mess of thoroughly mangled sediments and bedrock seafloor all scrambled together.

Beneath the veneer of basalt lava flows that covers the seafloor there is a zone of serpentinite some thousands of feet thick. Serpentinite is a strange, greasy-looking rock that comes in various shades of green and may be soft enough to carve with a knife. Found with it is a rock which many people call soapstone and which really does feel almost like soap. This light-colored rock is ground up to make talcum powder. Serpentinite forms when the heavy, black rocks of the earth's interior absorb seawater at the rift where new oceanic crust is created.

Serpentinite is not only soft but weak and slippery as well. When a slab of bedrock seafloor gets crammed into a growing coastal range, its serpentinite squeezes through every available weak spot in the accumulating mass as though it were an overripe banana in a bale of hay. Occasionally masses of serpentinite manage to slip back to the surface after having started their ride down into the earth's interior, bringing along chunks of basalt and even some sedimentary rock that had been carried down and briefly cooked under extremely high pressures that prevail many miles below the surface. These compressed rocks are called blueschists and eclogites and they are fairly common as lumps and locally large masses in some of the serpentinite bodies in the Klamaths. Most blueschists are rather darkly unattractive but a few blueschists are boldly flecked with big crystals of sky blue minerals and some eclogites are beautiful aggregates of red garnets and green jadeite; a few are actually jade. No other rocks are quite as tough and hard to break as blueschists and eclogites so they are much in demand for jettystone wherever they occur along the south coast of Oregon.

Meanwhile, most of the descending seafloor slides on down into the interior of the earth, finally getting hot enough that its basalt veneer melts and the serpentinite loses its water. The molten basalt and steam rise back to the surface where they fuel a chain of volcanoes parallel to the seaward edge of the growing coastal range and usually about 100 miles inland from it. The descending slab of seafloor will finally lose its identity and once again be part of the earth's interior when it has lost its covering of basalt, the serpentinite has lost its water, and the entire slab has heated up. This probably happens by the time it has sunk to a depth of about 200 miles.

Some of the basalt magma melted off the sinking seafloor rises directly to the surface where it erupts to make black lava flows. If the basalt rises through old continental rocks rich in silica it may melt and assimilate them to form large masses of lighter-colored magma enriched in silica — andesite. The silica-rich rocks deep in the continental crust have a much lower melting point than basalt and are almost hot enough to melt anyway so a relatively modest amount of basalt magma can cook up quite a batch of andesite. Much of the andesite magma manages to rise all the way to the surface and erupt through volcanoes but some of it crystallizes and cools underground to form large intrusions of granitic rock. Erosion eventually exposes these at the surface long after the volcanic activity has ended.

As the coastal range grows seaward, the line of seafloor sinking shifts seaward and so does the chain of volcanoes. Eventually the basalt magma rising from the sinking seafloor begins to get into the older part of the coastal range which also consists of silica-rich rocks which also assimilate into the basalt to form andesite magmas. The andesite magmas forming at the base of the coastal range rise upward through it, intruding and cooking the crumpled sedimentary rocks. This is the final step in the transformation of sedimentary rocks into new continental crust, the final step in adding a bit to the continent. The Blue, Wallowa, and Klamath Mountains are all composed of well-cooked and deformed sedimentary rocks and are full of large granitic intrusions; they are additions to the continental crust that have formed in Oregon within the last 200 million years.

When the collision between the North American continent and the Pacific Ocean floor began about 200 million years ago, the Blue and Wallowa Mountains, the Klamath Mountains, and the Sierra Nevada of California were all part of one continuous coastal range forming in the same way and at the same time. Then sometime between 150 and 100 million years ago a very strange thing happened to the Klamaths — they broke away from the rest of the chain and moved westward about 60 miles to become an island standing offshore from the coast.

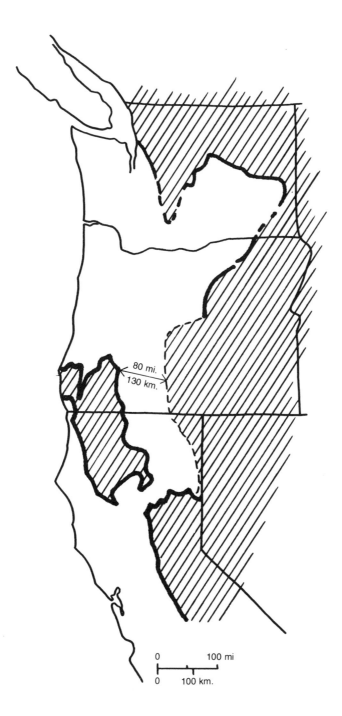

80 mi.
130 km.

0   100 mi
0   100 km.

Map showing Klamaths separated from the rest of the coastal range, a movement that had occurred by about 100 million years ago.

It is really quite unusual for a segment of a developing coastal range to detach itself from the continent and move offshore to become an island. We can't explain why or how it happened. But there is no doubt that it did indeed happen. The southern edge of the Klamaths and the northern edge of the Sierra Nevada in California are both just as sharp, as though they had been cut with a knife and the rocks would fit almost perfectly across the line of the cut if the two mountain blocks could somehow be slid back together. There is no doubt that they must once have been a single continuous range that was broken into separate pieces by movement along a fault. The northern edge of the Klamaths and the western edge of the Blue Mountains in Oregon are both buried under younger volcanic rocks so the picture is not so clear. But there are reasons, which we will discuss in later chapters, for believing that they are probably separated from each other.

Neither is there any room for doubt that the Klamaths really were an offshore island or at least a long peninsula. All along their eastern flank, from the area a few miles north of Medford to the southern end of the range in California, there are beds of sedimentary rocks full of fossil shells of animals known to have lived in seawater during late Cretaceous time, between about 100 and 60 million years ago. So we can be sure that an inland seaway separated the Klamaths from the coast and it is probably reasonable to assume that it was about 60 miles wide because that is the amount of the east-west separation between the northern Sierra Nevada and the Klamaths in California.

The late Cretaceous marine sedimentary rocks along the eastern flank of the Klamaths are partly buried beneath younger sedimentary rocks deposited on dry land during Eocene time, about 50 million years ago. Evidently the seaway separating the Klamath island from the coast had fairly well filled by then, converting the island to an outlying group of mountains connected to the mainland by a low plain.

Whatever it was that moved the Klamaths westward, separated them from all the volcanic activity as well as from the mainland. None of the big masses of granitic rock intruding the Klamaths are younger than about 150 million years old.

Younger granitic intrusions are abundant both north and south of the Klamaths in the Blue and Wallowa Mountains of Oregon and the Sierra Nevada of California. Of course the original volcanoes have long since eroded off all those ranges but we can infer their former presence from the big masses of granite.

The same pattern of collision between ocean floor and continent that began about 200 million years ago persisted until about 35 million years ago with the line of seafloor sinking and the chain of volcanoes shifting steadily seaward. Parts of the last chain of volcanoes associated with it are still visible in the Ochoco Mountains of central Oregon and in areas of the southern Cascades where we find volcanic rocks erupted about 30 million years ago.

Then in the middle of Oligocene time, about 35 million years ago, the pattern changed as the line of seafloor sinking abruptly shifted to essentially its present position offshore from the modern coast. The expanse of seafloor between the two lines of sinking stopped moving, the old coastal range stopped growing, and the old volcanic chain quit erupting. Meanwhile, a new coastal range began forming offshore and new volcanoes began erupting along a line parallel to the present Cascades. That shift in the line of seafloor sinking brought to an abrupt end the first and longest chapter in Oregon's geologic history and started the sequence of events that complete the story and add most of what we now see.

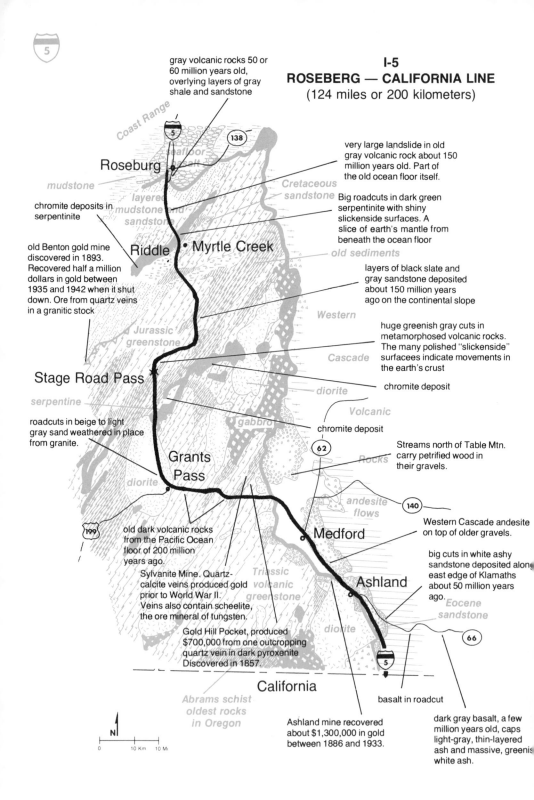

gray volcanic rocks 50 or 60 million years old, overlying layers of gray shale and sandstone

Coast Range

Roseburg

mudstone

layered mudstone and sandstone

chromite deposits in serpentinite

Cretaceous sandstone

very large landslide in old gray volcanic rock about 150 million years old. Part of the old ocean floor itself.

Big roadcuts in dark green serpentinite with shiny slickenside surfaces. A slice of earth's mantle from beneath the ocean floor

Riddle • **Myrtle Creek**

old sediments

old Benton gold mine discovered in 1893. Recovered half a million dollars in gold between 1935 and 1942 when it shut down. Ore from quartz veins in a granitic stock

layers of black slate and gray sandstone deposited about 150 million years ago on the continental slope

Western

Jurassic greenstone

huge greenish gray cuts in metamorphosed volcanic rocks. The many polished "slickenside" surfacees indicate movements in the earth's crust

Cascade

Stage Road Pass

serpentine

diorite

chromite deposit

Volcanic

roadcuts in beige to light gray sand weathered in place from granite.

gabbro

chromite deposit

diorite

Grants Pass

Streams north of Table Mtn. carry petrified wood in their gravels.

Rocks

andesite flows

old dark volcanic rocks from the Pacific Ocean floor of 200 million years ago.

Western Cascade andesite on top of older gravels.

Medford

big cuts in white ashy sandstone deposited along east edge of Klamaths about 50 million years ago.

Sylvanite Mine. Quartz-calcite veins produced gold prior to World War II. Veins also contain scheelite, the ore mineral of tungsten.

Triassic volcanic greenstone

Ashland

Eocene sandstone

Gold Hill Pocket, produced $700,000 from one outcropping quartz vein in dark pyroxenite Discovered in 1857.

diorite

California

Abrams schist oldest rocks in Oregon

basalt in roadcut

Ashland mine recovered about $1,300,000 in gold between 1886 and 1933.

dark gray basalt, a few million years old, caps light-gray, thin-layered ash and massive, greenish white ash.

N

0   10 Km   10 Mi

the klamaths

interstate 5

roseburg — california line

The geologic plot thickens about 5 miles south of Roseburg, at the Winston turnoff, where the interstate highway leaves the relatively simple rocks of the Coast Range and enters an area of older and much more complex rocks. These are the northernmost extension of the Franciscan rocks of the California Coast Range. Then a few miles farther south, in the vicinity of Stage Road Pass, the road crosses the much less distinct boundary that separates these from the still older and even more complex rocks of the Klamath Mountains which are basically similar except that they also contain masses of granitic rock.

All these older rocks started out, like the younger ones farther north, as sediments deposited on the seafloor and then scraped off against the continent. The granites rose as molten masses of andesitic magma, into the older rocks while the moving seafloor was scraping the younger ones onto the accumulating pile.

Most of the rocks between Winston junction and Stage Road Pass are volcanics that were sheared, altered and recrystallized almost beyond recognition as the moving seafloor jammed them into the edge of the continent. There are some old seafloor sediments, also sheared, along a few miles of road just south of Winston and again in the immediate vicinity of Myrtle Creek.

The crushed and altered volcanic rock between Winston and Stage Road Pass has the chemical composition of andesite, not basalt. Andesite almost always erupts from big volcanic chains like the Cascades and never forms part of the seafloor. So it seems likely that

a chain of andesite volcanoes stood near here back in Jurassic time, about 150 million years ago, when all these rocks were formed and then rudely shoved into the continental margin. Maybe those old volcanoes looked something like the modern Aleutians which are andesitic and mark a line of encounter between the moving floor of the Pacific Ocean and another crustal plate.

About 2 miles north of Myrtle Creek and again at Stage Road Pass, the highway crosses belts of serpentinite, a rock that most certainly began as part of the oceanic crust. Serpentinite is a nasty-looking greenish rock always fractured into small chunks which have polished surfaces. Some serpentinite is soft enough to carve with a pocketknife. In places it is greenish-white, feels soapy, and many people call it soapstone. Serpentinite forms in the lower part of the bedrock seafloor; the belts of it in the older Coast Range mark the boundaries of slices of seafloor added to the continent.

Shiny slip surfaces in greenish serpentinite develop as the rock deforms under pressure.

Belts of serpentinite are easy to see because the rock weathers to an impoverished orange soil on which few plants will grow. So serpentinite is always thinly cloaked in a cover of distinctively scrawny bushes and struggling trees with very little grass growing between them. In some places the orange soil and in others the greenish bedrock shows between the plants. No place to start a vegetable garden.

The only nickel mine in the country operates in one of those belts of serpentinite at Nickel Mountain just west of Riddle. The deposit was discovered back in 1865 but serious mining didn't begin until 1954 although several attempts failed in the meantime. The Riddle mine now produces about 25 million pounds of refined nickel each year.

Serpentinite normally contains a small amount of nickel which may concentrate in the soil if conditions are just right. It is necessary to start with one of the harder varieties of serpentinite and to weather it under very wet climatic conditions so that a red lateritic soil forms. Nickel, like iron, is retained in such soils and if weathering reduces a thick enough section of rock to laterite, the nickel may accumulate to a mineable concentration. The mine at Riddle works such soil that runs about 1.4 per cent nickel. Mining the deposit is essentially a matter of stripping the nickel-bearing soil off the top of the mountain.

Southwestern Oregon contains a lot of serpentinite and a lot of red soil and it might seem that there should be plenty of other deposits like the one on Nickel Mountain near Riddle. But the perfect combination of serpentinite at just the right stage of alteration and the right degree of weathering seems to be a tricky one and so far no other large deposits have turned up.

Almost every belt of serpentinite contains small amounts of chromite, a heavy black mineral which is the only source of chromium. Because it is such a greasily mobile rock, most serpentinite is intensely deformed and the chromite is strung through it like streaks of fat marbled through a steak. It is easy enough to find small amounts of chromite in the serpentinite belts but very difficult to locate enough in one place to make a mine. Nevertheless, about a dozen small chromite mines have operated from time to time in the serpentinite belts near Interstate 5. One group of mines is in the Riddle area and another a few miles northeast of Stage Road Pass. This country imports most of its chromite from east Africa and none of the Oregon deposits are competitive except when the overseas sources are cut off

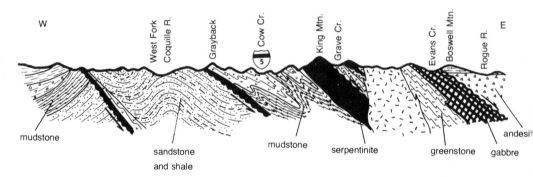

by war or other political problems. Oregon produced chromite during both world wars and again during a government strategic stockpiling program in the early 1950's.

Grants Pass is near the eastern edge of a large mass of dark granitic rock which is not well exposed near the road. The route between Grants Pass and Gold Hill generally follows the Rogue River through hills eroded into rocks that began as sediments and volcanics laid down at least as long as 200 million years ago, during Triassic time. About 200 million years ago the moving seafloor shoved the whole mess into the edge of the growing continent and then later, as more rocks jammed in on top of them, the rocks in this part of the Klamaths got hot enough to cook and were intruded by large masses of andesitic magma, some of which cooled underground to make coarsely crystalline granitic intrusions. Now the older rocks are so transformed that it is hard to figure out what they may have been originally. Most of the numerous large roadcuts along the route between Grants Pass and Gold Hill expose very dark rocks that probably started out as volcanics.

Between the area about 5 miles north of Medford and the California line, Interstate 5 skirts the eastern edge of the Klamaths, tracing the boundary between them and the western Cascades. All the rocks in the mountains west of the road are metamorphosed sediments and volcanics intruded by granites, all of them at least 150 million years old. Rocks in the hills east of the highway are dark volcanics, andesite and basalt, erupted when the western Cascade volcanic chain was active.

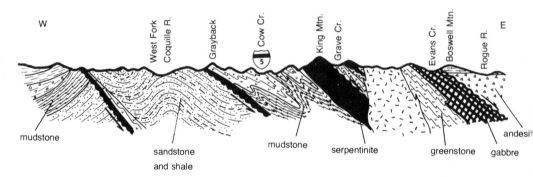

Section across the line of Interstate 5 near Medford.

Just what happens at the eastern edge of the Klamaths is a difficult question. Do the Klamaths continue northeastward buried beneath younger volcanic rocks of the Cascades and central Oregon to reappear in the geologically similar Blue Mountains or does some kind of gap filled by other rocks separate these two ranges? The belt of rocks along Interstate 5 between the area north of Medford and the California line provides the basis for some interesting conjecture.

The important exposures aren't much to look at: soft sediments barely hard enough to pass for rock and rarely resistant enough to make good outcrops. Patches of these sediments in the hills west of the road next to the older rocks of the Klamaths contain fossils of animals that lived in seawater during late Cretaceous time, perhaps 75 million years ago. The rocks along the road are brown sandstones, well exposed in roadcuts about 5 miles south of Medford, that contain fossil leaves of trees that lived during Eocene time, about 45 million years ago. These formations are deposits laid down along the coast of an inland sea that flooded northeastern California and south-central Oregon during late Cretaceous time. Apparently the Klamaths were

Flat caps on a few sandstone hills are all that remains of a basalt flow that poured across the area several million years ago. Erosion has since removed most of the flow as it carved the modern landscape. Scene a few miles north of Medford.

then an island or a peninsula separated from the similar rocks in the Blue Mountains by seawater. It seems that the Klamaths really do end along the line of Interstate 5 and do not continue eastward buried beneath younger rocks.

Between this area and that a few miles north of Medford and Ashland, the rocks exposed in the Klamath Mountains west of the road are metamorphosed sediments and volcanics intruded by a few small masses of granite, basically similar to the rocks along the road between Grants Pass and Gold Hill. Mountains west of the road between Ashland and the California line are eroded into a single large mass of granite.

Oregon's oldest known rocks are in the Siskiyou mountains right along the state line about 20 miles west of Interstate 5. They are thoroughly cooked and recrystallized sediments thought to be at least 425 million years old. Unfortunately they are in a nearly inaccessible region of very rugged mountains and there is no easy way to see them.

Layers of muddy sandstone that accumulated on the Pacific Ocean floor and now stand tilted at a sharp angle. Exposed near the Oakland exit from I-5, north of Roseburg.

u.s. 101

port orford — california line

Most of the route between Port Orford and the California line clings precariously to a continuously rocky coastline with mountains rising dramatically from the surf. The road is almost continuously within sight of the ocean and never beyond the sound of pounding surf.

Rocks along this southernmost stretch of the Oregon coast are distinctly different from those farther north. From the geologic point of view, they are a thin panhandle of the northern California Coast Range extending northward into Oregon. Actually, they are hard to distinguish from the rocks in the Klamath Mountains.

These southern Coast Range rocks began, like those farther north, as sediments laid on the floor of the Pacific Ocean. But these rocks, unlike those farther north, have actually been scraped off the seafloor onto the edge of the continent as the oceanic crust slid out from under them into an ocean trench. So these rocks are severely deformed and recrystallized enough to be very hard; they are no longer soft mudstones and sandstones. They began as sediments laid on the seafloor back in Jurassic and Cretaceous time, between 200 and 100 million years ago in round numbers, and then jammed into the Coast Range during that same time interval. All available evidence suggests that the Pacific Ocean floor moved very rapidly during that time and the chaotically jumbled rocks in the southern Coast Range certainly appear to have been roughly handled.

The older rocks in the southern Coast Range also contain serpentinite, the dark green, rather greasy looking rock which forms part of the oceanic crust. It is as greasy as it looks and some of it manages to squirt back to the surface through the harder rocks as though it were ketchup squeezing out of a hamburger. It usually works its way through zones of broken rocks so wherever it occurs we suspect it may mark a fault.

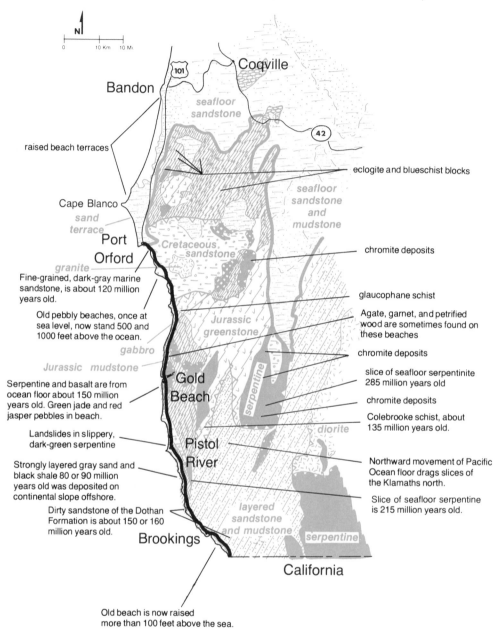

raised beach terraces

Coqville

Bandon

seafloor sandstone

eclogite and blueschist blocks

Cape Blanco

sand terrace

seafloor sandstone and mudstone

Port Orford

Cretaceous sandstone

chromite deposits

granite

Fine-grained, dark-gray marine sandstone, is about 120 million years old.

glaucophane schist

Old pebbly beaches, once at sea level, now stand 500 and 1000 feet above the ocean.

Jurassic greenstone

Agate, garnet, and petrified wood are sometimes found on these beaches

gabbro

chromite deposits

Jurassic mudstone

slice of seafloor serpentinite 285 million years old

serpentine

Serpentine and basalt are from ocean floor about 150 million years old. Green jade and red jasper pebbles in beach.

Gold Beach

chromite deposits

Colebrooke schist, about 135 million years old.

Landslides in slippery, dark-green serpentine

diorite

Pistol River

Strongly layered gray sand and black shale 80 or 90 million years old was deposited on continental slope offshore.

Northward movement of Pacific Ocean floor drags slices of the Klamaths north.

Slice of seafloor serpentine is 215 million years old.

Dirty sandstone of the Dothan Formation is about 150 or 160 million years old.

layered sandstone and mudstone

Brookings

serpentine

California

Old beach is now raised more than 100 feet above the sea.

N

0   10 Km   10 Mi

34

Wherever there are serpentinites there is likely to be jade. It is almost impossible to find in the outcrop because it is the same color as serpentinite but much easier to find in river and beach gravels. Jade is a tough rock that easily survives the wear and tear of transport while the soft serpentinite is quickly ground to powder. A number of southern Oregon beach and river gravels contain jade pebbles.

The older rocks actually extend more than 20 miles north of Port Orford and are well exposed in the sea cliffs for half of that distance, but the road in that area is several miles inland on a broad marine terrace with very few outcrops so most people won't see them unless they take the side road to Newburgh State Park. Between Port Orford and the California line the older Coast Range rocks are exposed almost continuously along the road. The vast majority of them are dirty sandstones now hardened by recrystallization into very solid rock which is usually rather dark and massive looking. But there are other kinds of rocks.

Humbug Mountain, about 6 miles south of Port Orford, is a huge mass of gravelly conglomerate deposited during Cretaceous time, a little more than 100 million years ago. The fact that the original deposit was mostly gravel suggests that it must have been laid down near shore because pebbles rarely get very far out on the seafloor.

The seaward side of Humbug Mountain descends right into deep water with no wave-cut shelf on its face wide enough for a roadway. The highway cuts inland a short ways and follows a beautifully forested creek valley around the eastern side of the mountain. It seems odd but is true that very steep cliffs may be almost immune to wave erosion if they descend directly into deep water with no shoal at

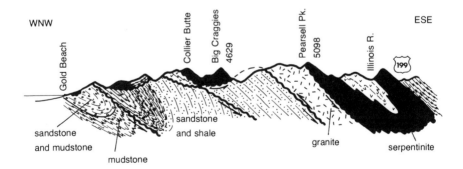

Section across the Klamaths in the vicinity of Gold Beach.

Good drainage is the best safeguard against landslides. This perforated pipe helps stabilize a roadcut in serpentinite about 2 miles north of Gold Beach.

their base. Waves pile into surf and then break because they drag on a shallow bottom. If they hit a cliff before they feel bottom, they will not break but instead slosh harmlessly up and down its face, just as they do against the side of a boat in deep water. The deep water immediately offshore from Humbug Mountain is protecting its face from wave erosion.

Gold Beach is within a big belt of serpentinite spectacularly exposed in several roadcuts both north and south of town. The rock is unmistakeable — green and sheared into minute slivers but containing solid chunks which have rounded outlines and brightly polished surfaces. These chunks appear to have slipped through the sheared groundmass as it moved as though they were so many watermelon seeds. Some have such fascinating shapes and colors that they qualify as natural sculpture.

Rainbow Rock, just south of Boardman State Park, is composed of intricately folded thin beds of colorful chert, a very hard sedimentary rock composed almost entirely of silica. There are quite a few such bedded cherts in this part of the Coast Range but Rainbow Rock is one of the few places where it is easy to get a good look at them.

This kind of bedded chert forms far out on the ocean floor in deep water, beyond the reach of sands and muds washed in from the continent. The only sediments that settle on those remote reaches of the seafloor are the microscopic skeletons of one-celled animals called

Sea stacks made of resistant masses of rock stand against the surf near Gold Beach.

radiolaria and plants called diatoms, both of which are made of silica. Other kinds of shells dissolve in those deep, cold waters. The tiny shells accumulate very slowly to form an ooze with a consistency about like toothpaste which eventually hardens by recrystallization into chert. So what we see at Rainbow Rock are deep-sea oozes that almost certainly accumulated many hundreds of miles offshore and finally made it into the Coast Range where they are mixed with sandstones that were deposited much closer to shore.

Just north of Brookings the road gets onto a broad marine terrace planed to a smooth, gently-sloping surface by wave action when it was slightly below sea level and now raised about 50 feet above the waves. It is very difficult to establish the age of a marine terrace, especially on a coastline as active and unstable as Oregon's, but a good guess for this one might be in the neighborhood of several hundred thousand years. The occasional big rocks that stand prominently above the terrace surface are old sea stacks.

The wrinkled hill and rough road are both due to a large landslide. A common sight in the Coast Range, this is one beside U.S. 101 about 6 miles south of the Pistol River.

u.s. 199

grants pass — california line

The route through the rugged mountains between Grants Pass and Crescent City passes some of the nicest scenery and most interesting rocks in the northwest. Scenically these are the Siskiyou Mountains; geologically they are the Klamaths. Good scenery and interesting rocks usually go together and these mountains are no exception.

The Klamaths are a tangled geologic puzzle of chaotically scrambled rocks that began on the Pacific Ocean floor sometime between 200 and 100 million years ago. The sinking seafloor scraped them off onto the edge of the continent during that same time interval and then large masses of molten andesitic magma intruded the crumpled rocks, some of it erupting through volcanoes and the rest crystallizing underground to form granitic masses. Such complex mountains are hard to figure out even where the bedrock is well exposed; the dense forests of the Klamaths make the puzzle almost impenetrable. The combination of wild rocks and poor exposure is hard to crack.

Grants Pass is in a large mass of dark granitic rock, in this case a rock technically known as diorite. It is an oval body which measures about 17 miles from north to south and about 8 miles from east to west. U.S. 199 crosses its width between Grants Pass and Wilderville as it follows the Rogue River valley.

All the way between Wilderville and the state line the highway stays on the outcrop of the Galice formation, a thick sequence of dark mudstones deposited on the floor of the Pacific Ocean some 150 million or so years ago and scraped off into the Klamaths not long thereafter. Along most of this distance, all the way from Selma to the state line, the highway stays within a mile or so of the eastern edge of a large slab of bedrock seafloor — not basalt lava flows but the dark serpentinites and peridotites that lie beneath, really a slice off the

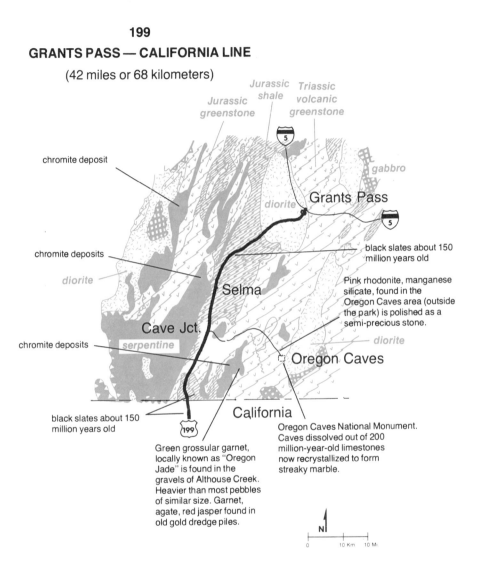

199
GRANTS PASS — CALIFORNIA LINE
(42 miles or 68 kilometers)

Jurassic greenstone

Jurassic shale

Triassic volcanic greenstone

chromite deposit

gabbro

diorite

Grants Pass

chromite deposits

black slates about 150 million years old

diorite

Pink rhodonite, manganese silicate, found in the Oregon Caves area (outside the park) is polished as a semi-precious stone.

Selma

Cave Jct.

serpentine

diorite

chromite deposits

Oregon Caves

black slates about 150 million years old

California

Green grossular garnet, locally known as "Oregon Jade" is found in the gravels of Althouse Creek. Heavier than most pebbles of similar size. Garnet, agate, red jasper found in old gold dredge piles.

Oregon Caves National Monument. Caves dissolved out of 200 million-year-old limestones now recrystallized to form streaky marble.

N

0   10 Km   10 Mi.

earth's interior. Such slabs sometimes get mixed into the mountains growing next to the line of seafloor sinking, instead of sliding peacefully back into the earth's interior as they should.

Those big slabs of heavy, dark rock that really belong to the earth's interior usually have some interesting minerals in them, things we don't see in ordinary continental rocks. They always contain some platinum, never very much but then no other kind of rock contains any at all. There is never enough platinum to make a hard rock mine pay but streams that drain such areas always contain a bit in their gravels. Small platinum placers are common in the Klamaths but none are big enough or rich enough to be worth mining on a large scale. Chromium and nickel also occur in such rocks and the Klamaths are full of chromium and nickel prospects at least some of which will probably become mines someday.

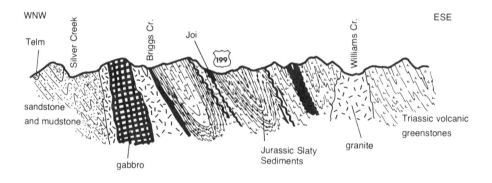

Section across the line of U.S. 199 just north of the state line.

Oregon Caves National Monument is 18 miles east of the highway on an excellent paved road. The caves are in marble which formed from folded limestone when a nearby mass of igneous rock heated and recrystallized it. Good exposures of intricately folded marble add interest to the walk from the parking lot to the cave entrance. Limestone and marble are very similar rocks; both are made of the mineral calcite and the only important difference between them is that marble contains larger crystal grains. Both rocks dissolve in weak acids and ordinary rainwater is acid enough to attack them. Rain seeping into the rock along fractures slowly dissolves holes to create caves. Then more water dripping into the caves evaporates and deposits part of its load of dissolved calcite to create the elegant dripstone formations that make them so beautiful.

*A lush growth of
fat stalactites
in Oregon Caves.*

The rock at Oregon Caves began as limestone deposited in sea water some 200 million or so years ago. By about 150 million years ago it had been squashed into the Klamaths and transformed into marble by recrystallization. It is hard to know how old the caves may be. Caves are hard to date because they rarely contain much internal evidence of their age but when they do, it rarely suggests that they are much older than about a million years. The dripstone formations are certainly much younger than that. Again, no one knows how old they are but dripstone can form fairly quickly under the right conditions. Stalactites can easily grow an inch every decade if a steady drip of water keeps adding to them.

Stalactites hang from the ceiling and a sumptuous flow of dripstone upholsters the floor of a room in Oregon Caves.

The craggy peaks of Elkhorn Ridge make a dramatic background west of the Baker Valley.

the blue mountains

interstate 80

baker — ontario

Baker is at the south end of a broad, flat valley within the Blue Mountains. It nestles at the east end of Elkhorn Ridge which contains rocks originally deposited as sediments on the continental shelf about 300 million years ago and then squashed into the Blue Mountains when the seafloor began sinking in this area approximately 100 million years later.

Many of the creeks in Elkhorn Ridge contained lavish deposits of placer gold. The old timers made it their first priority to clean them out back in the last century but never could find bedrock lode deposits large enough to keep things going. It seems that the bedrock in Elkhorn Ridge contains plenty of gold but it is all disseminated through many widely scattered small stringers of quartz instead of in a few large veins which could be mined.

It is true, as the old timers say, that "Gold is where you find it," but it is also true that where you find gold is usually near the margins of granite intrusions. Early prospectors found bedrock lode deposits of gold around the granites in the Blue Mountains about 1861 and by the following year they had swarmed up every creek and over every hill looking for more. We won't speculate as to whether or not they

BAKER — ONTARIO

(72 miles or 116 kilometers)

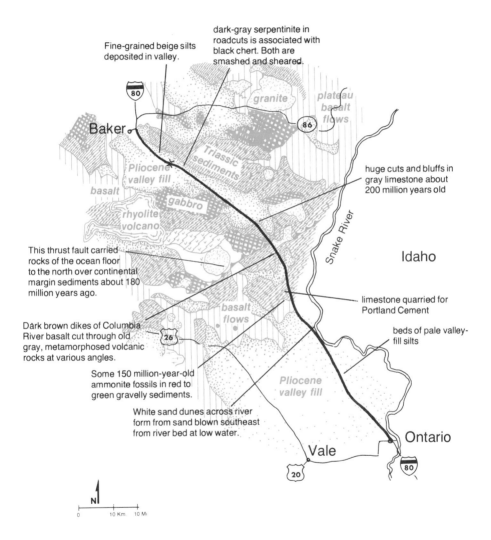

dark-gray serpentinite in roadcuts is associated with black chert. Both are smashed and sheared.

Fine-grained beige silts deposited in valley.

granite

plateau basalt flows

Baker

Triassic sediments

Pliocene valley fill

basalt

gabbro

rhyolite volcano

huge cuts and bluffs in gray limestone about 200 million years old

Snake River

Idaho

This thrust fault carried rocks of the ocean floor to the north over continental margin sediments about 180 million years ago.

limestone quarried for Portland Cement

basalt flows

beds of pale valley-fill silts

Dark brown dikes of Columbia River basalt cut through old gray, metamorphosed volcanic rocks at various angles.

Some 150 million-year-old ammonite fossils in red to green gravelly sediments.

Pliocene valley fill

White sand dunes across river form from sand blown southeast from river bed at low water.

Ontario

Vale

N

0   10 Km.   10 Mi.

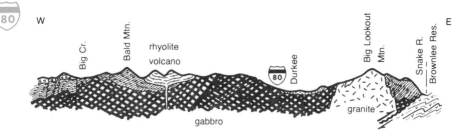

Section across the line of I-80 near Durkee.

found Oregon's legendary "Blue Bucket" mine which they all sought. Within a few years little mining towns had sprung up like mushrooms after a spring rain and many thousands of people were living and working in parts of the Blue Mountains that are unpopulated today.

The earliest stages of gold mining are the best because the first arrivals get to skim the placer deposits in the creeks and mine the upper parts of vein deposits enriched by millions of years of weathering. Then comes the harder and less rewarding work of mining the deep stream gravels and the leaner primary ores in the bedrock veins. The first flush of fast and easy profits from gold in the Blue Mountains was in and squandered by the time the Civil War had ended and what was left was decades of backbreaking work for low wages.

Most of the Blue Mountain lode deposits were lean and irregular. Only a very few mines produced more than a million dollars worth of bullion and quite a few of them never made enough to pay off the original investment. Rising operating costs and fixed gold prices put them all in an economic squeeze that closed one mine after another during the decades after the turn of the century and the people slowly filtered out of the hills looking for other ways to make a living. All the surviving mines were closed at once by a government order issued in 1942 because gold mining was diverting essential manpower from other industries more important to the war effort. Very few of the Blue Mountain gold mines even attempted to reopen after 1945 and the only one that shipped a significant tonnage of ore was the Buffalo Mine which is in the hills about 8 miles north of the ghost town of Granite. It was the most persistent mine in the district even if not the most productive.

A number of the mines had developed reserves ready to mine when they closed. Mining stopped more often for economic reasons than for lack of gold-bearing rock and many mines will reopen if the price of gold, which is no longer fixed, gets high enough to enable them to operate profitably. Most people will be happier if that never happens because gold mining flourishes during hard times of high unemployment and low labor costs.

The modern highway follows close to the path of the old Oregon Trail between Baker and Ontario, passing through the Blue Mountains in a series of broad valleys opened as blocks of the crust dropped along a swarm of faults. They are floored by Pliocene valley fill gravels. The hills on either side of the road between Baker and the area just north of Olds Ferry on the Snake River belong to the Blue Mountains. Roadcuts beside the highway are in old sedimentary rocks first laid down on the continental shelf and slope between 300 and 200 million years ago and then shoved together into a crumpled mass when the sinking seafloor scraped them off against the edge of the continent. Some of the hills also contain gabbro, a black igneous rock which must have been part of the oceanic crust itself and somehow managed to get scrambled into the mountains instead of sliding back into the earth's interior. And there are intrusions of granitic rock that penetrated the whole mess as molten magma.

Flood basalt flows beside Interstate 80 just north of Huntington.

Between Olds Ferry and Ontario the road crosses a broad expanse of valley-fill sediments laid on top of basalt flows during the Pliocene episode of arid climatic conditions. A small patch of these flows is exposed near the road just north of Huntington. This was also the area in which the flood basalts impounded a huge lake which extended east as far as Boise during Miocene time, about 15 or 20 million years ago. It was almost certainly the overflow spillway of this Miocene lake which established the present course of the Snake River downstream through Hell's Canyon. There are nice exposures of the sediments deposited in that lake in the area between the interstate highway and U.S. 26 but none anywhere near either road. They contain quite a variety of distinctly tropical fossil leaves.

Basalt lava flows.

oregon 86

baker — hell's canyon

The road between Baker and Hell's Canyon crosses older bedrock belonging to the Blue Mountains, younger lava flows that belong to the Columbia Plateau, and still younger valley-fill sediments. But the main geologic attraction of the drive and reason enough for the effort is Hell's Canyon itself, the deepest gorge in the country. There the Snake River has cut through the thick crust of basalt and deep into the older rocks of the Blue Mountains, exposing a magnificent geologic cross-section.

The road crosses about a mile of brownish flood basalt right at the eastern edge of the Baker Valley. This is part of the big area of flows that caps most of the Farley Hills which fringe the eastern margin of the valley.

Between the basalt and the area about 3 miles west of Richland, the road crosses large areas of gabbro, a coarsely granular and rather forbiddingly black rock composed of greenish-white crystals of plagioclase set in a matrix of black pyroxene. Younger intrusions of a dark granitic rock called diorite cut the gabbro and the road crosses these too. The Powder River has cut a picturesque little canyon through these rocks, creating wonderful outcrops. Most of the gabbro is pretty streaky looking stuff which means that the stresses of mountain building squeezed and stretched it along with the other rocks in the area. It was almost certainly a part of the oceanic crust which got mixed into the sedimentary rocks of the Blue Mountains. There is a lot of gabbro in the oceanic crust.

Big patches of soft valley-fill sediments floor the low places between the areas of bedrock outcrop. These sediments are a mixture of gravel, mud, and colorful beds of pale volcanic ash. They are debris washed into the low places in the landscape during the Pliocene

86

BAKER — HELLS CANYON DAM

(95 miles or 153 kilometers)

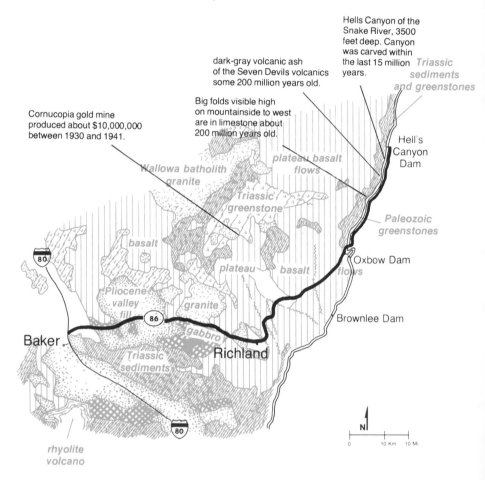

Hells Canyon of the
Snake River, 3500
feet deep. Canyon
was carved within
the last 15 million   *Triassic*
years.   *sediments*
                       and greenstones

dark-gray volcanic ash
of the Seven Devils volcanics
some 200 million years old.

Big folds visible high
on mountainside to west
are in limestone about
200 million years old.

Hell's
Canyon
Dam

Cornucopia gold mine
produced about $10,000,000
between 1930 and 1941.

*plateau basalt
flows*

*Wallowa batholith
granite*

*Triassic
greenstone*

*Paleozoic
greenstones*

basalt

Oxbow Dam

plateau *basalt* *flows*

*Pliocene
valley
fill*

granite

Brownlee Dam

Baker

gabbro

Richland

*Triassic
sediments*

N

0   10 Km   10 Mi.

*rhyolite
volcano*

Platy layering in basalt beside the Powder River near Halfway. This seems to develop where basalt cools on a slope and continues to flow as it solidifies.

period of dry climate when the plant cover was too sparse to protect the ground against erosion and the streams didn't carry enough water to wash the eroded material away.

The road between Baker and Richland passes through the Virtue mining district which produced between 2 and 3 million dollars in gold during the period from the start of the Civil War until the turn of the century. The early prospectors found placer deposits in a gulch and traced the gold to its source thus locating the Virtue Mine which produced from a swarm of quartz veins in a fault zone in the gabbro several miles south of the road. The Flagstaff Mine, which is about a half mile north of the road, never produced very much and the other mines and prospects in the district all flopped.

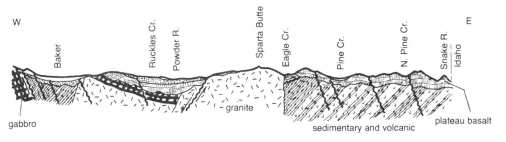

Cross-section from Baker through Hell's Canyon showing the older rocks of the Blue Mountain province buried beneath the plateau basalt.

49

Except for another patch of valley-fill sediments in the area around Halfway, all the rocks between Richland and the mouth of Hell's Canyon are flood-basalt lava flows erupted from the Grande Ronde volcano in the northeastern corner of Oregon during Miocene time, about 15 or so million years ago. These old lava flows are rather well covered by fertile soil.

The Wallowa Mountains in the distance north of the road began as a mass of continental-shelf and ocean floor sediments scraped off onto the edge of the continent about 200 million years ago, quite early in the sequence of geologic events that made Oregon. A large mass of granite intruded those rocks about 150 million years ago and now makes the resistant core of the higher parts of the range.

Weathered basalt near Homestead.

Crumpled layers of sedimentary rock exposed in the walls of Hell's Canyon.

Granite intrusions generally spawn ore bodies around their margins and those in the Wallowa Mountains conformed to tradition. The Cornucopia mining district in the forested mountains north of Halfway began producing gold in 1895. A cyanide mill revived the district in 1912 and kept it going intermittently for several decades. Like so many old mining camps, the ghost town of Cornucopia is near the contact between a mass of granite and the older rocks that surround it.

Nothing remains now of Copperfield and Homestead but old mine dumps and names lightly printed on road maps. Prospectors found ore-bearing quartz veins cutting through old sedimentary rocks in the hills west of the Snake River in the 1890's and mining began in 1910 after a long period of development work. During the first World War the mines and mill went full blast and these were flourishing little places tucked away in the mouth of Hell's Canyon where hardly anyone outside might notice them. Then the law cracked down and a few years later the ore got lean and before long both the fun and the money were gone. The mines closed as the depression began after having produced about 15 million pounds of copper, over a quarter of a million ounces of silver, and nearly a million dollars worth of gold. Not a bad record for a couple of places that don't even exist anymore.

Looking north into the south end of Hell's Canyon.

Hell's Canyon is simply fantastic. The Snake River has cut its gorge down through the lava plateau and into the contorted rocks of the Blue Mountain province buried beneath it. The upper canyon walls are several thousand feet of basalt ledges stained brown by weathering. The lower part is in old sedimentary rocks that once formed a broad continental shelf before the moving sea floor telescoped them together into the edge of the continent about 200 million years ago to make the Blue Mountains.

The geologic events that created the rocks of the Blue Mountains had been over for almost 150 million years and erosion had carved them into a rugged landscape when floods of basalt from the Grand Ronde volcano buried all but the highest ranges, such as the Wallowas. That happened during the middle to latter part of Miocene time, probably about 15 or 50 million years ago.

The same lava flows that buried the mountains also impounded a big lake in the Boise region. It seems almost certain that overflow from that lake cut a spillway which became the course of the Snake River through what is now Hell's Canyon. We can be absolutely certain that the river has cut the entire canyon since the last flood basalts erupted close to 13 million years ago.

the coast range

and willamette valley

The stage was set for the formation of the Coast Range and Willamette Valley about 35 million years ago when the line of seafloor sinking jumped from its old course curving inland to its present position offshore. A large slab of seafloor that had been moving in towards the continent was isolated between the old and new lines of sinking and quit moving to become the stationary platform on which most of northwestern and western Oregon is built. The Coast Range is simply the uplifted western edge of that slab of seafloor. Had the line of sinking not jumped, that slab would long since have sunk into the earth.

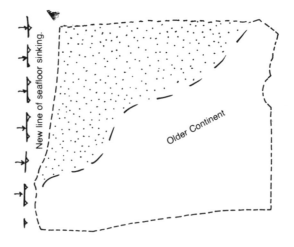

The stippled area shows the slab of seafloor that stopped moving about 35 million years ago when it was isolated between the old and new lines of seafloor sinking. Map is distorted to compensate for later movements.

Odd as it may seem, it is the sinking seafloor offshore that raised the Coast Range above sea level. As the sinking seafloor slides past the broken edge of the old slab that is the Coast Range, it scrapes off its load of muddy sediments, stuffing some of them under the edge of the old slab thus jacking it up. The Coast Range will rise as long as this continues.

When the seafloor finally quits sinking along its present line, the part of the Coast Range now offshore will rise above sea-level exposing the rocks now being stuffed under the edge of the Coast Range slab. Those will be the chaotically confused kind of rocks we see today in the Blue, Wallowa, and Klamath Mountains and in the Coast Range of California.

Except for its deformed southern end, the Coast Range is essentially an intact slab of seafloor. From a geologic point of view, studying the rocks in the Coast Range is the equivalent of taking an expedition to the bottom of the ocean. Better, actually, because streams have cut deep valleys into the Coast Range exposing its internal anatomy. All we can see of the modern seafloor is its upper surface.

Seafloor forms at rifts where two crustal plates are pulling away from each other. Basalt magma rises into the opening crack, fills it to make a dike, and pours out over the nearby seafloor as a lava flow. Basalt flows erupted under water look entirely different from those that erupt on land; they form a mass of bulging, lumpy forms that suggest a pile of oversized sofa pillows. An advancing lobe of lava chills quickly in contact with the cold sea water forming an outer skin of solid rock that bursts and lets the lava run out to do the same thing over again. The whole process is similar to the behavior of dripping candle wax and the forms produced are similar too. As flow after flow erupts from the rift that keeps opening between the separating plates, they pile up on each other to depths of hundreds or even thousands of feet to make the uppermost layer of hard bedrock beneath the seafloor. Meanwhile, the newly created seafloor is slowly moving away from the rift toward the continent and eventually to a zone of seafloor sinking. In the Coast Range we have a generous sample of seafloor that escaped its predestined doom of sliding back into the interior of the earth whence it came.

Of course the seafloor constantly receives sediments eroded from the continent and slowly carries them back to the continent and stuffs them into a coastal range. Were it not for the fact that new seafloor constantly forms while old seafloor constantly scrapes its sediments back onto the continent, the oceans would long since have filled with mud and the continents eroded to featureless plains. It is this constant creation and destruction of the restlessly moving seafloor that keeps our planet in business.

If the rift where the seafloor forms is not too far offshore, the sediments washing in from the continent will reach it and bury some of the lava flows almost as soon as they form. The result will be interlayered sediments and lava flows, something we often see in the Oregon Coast Range. Sometimes the molten basalt may not erupt onto the seafloor as a lava flow but instead inject itself between beds of sediment to make a sill — a layer of basalt sandwiched between layers of sediment. Some parts of the Coast Range are full of sills, many of them as much as several hundred feet thick.

The lava flows in the Coast Range erupted during Eocene time, perhaps 50 or 60 million years ago, so that must have been when this expanse of seafloor formed. Most of the dirty sandstones and mudstones that are interlayered in the seafloor lava flows and also deeply cover them in many parts of the Coast Range also date from Eocene time. Some parts of the Coast Range still contain younger Oligocene sediments deposited about 35 or 50 million years ago which presumably once blanketed the entire surface before erosion removed most of them.

It was near the middle of Oligocene time, about 35 million years ago, when the line of seafloor sinking jumped from its old course to its present one off the modern coastline. Evidently most of the Coast Range was jacked above sealevel very shortly thereafter because we find no sedimentary rocks deposited since then anywhere except in the Astoria and Tillamook areas. There we find mudstones deposited during Miocene time, some perhaps as recently as 20 million years ago, so evidently those areas remained submerged a few million years longer. Ever since middle Oligocene time, the seafloor sinking offshore has been jamming younger sedimentary rocks against and beneath the edge of the Coast Range slab.

Those Miocene mudstones in the northern end of the Oregon Coast Range contain a large variety of distinctly tropical fossil seashells nearly identical to those of the same age in southernmost California. Obviously the water along the Oregon coast was much warmer during Miocene time than it is now. Those seashells lived while red tropical soils were forming over most of Oregon and leaves of tropical plants were being preserved in eastern Oregon lake beds. Oregon was truly tropical during Miocene time.

In most of the Coast Range, the older sedimentary rocks are the kind of dark gray mudstones and dirty sandstones that accumulate on deep-sea floor far from shore, the same kind of sediments that come up in most deep-sea cores cut by oceanographic research vessels. But things are different at the southern end of the Coast Range, in the vicinity of Coos Bay and Coquille. There the sedimentary pile is very thick, tightly crumpled, and contains rocks deposited along the shoreline as well as some that appear to have been laid down in deep water.

The region around Coos Bay and Coquille is at the narrow southern tip of the big slab of seafloor isolated between the old and new lines of seafloor sinking. And it is also adjacent to the Klamath Mountains. It is easy to imagine that the Klamaths may have been a source of abundant sediment back in Eocene time which would have dumped onto the nearby seafloor to build a large continental shelf and coastal plain. Along that seashore there must have been big coastal marshes filled with lush jungles because we find thick coal seams in the area today. It is probably quite reasonable to imagine genuinely tropical jungles growing along the Oregon coast then because there is abundant evidence in many parts of the world to indicate that most of Eocene time, like Miocene time, was a period of un-usual warmth and wetness.

So imagine a large coastal plain and continental shelf built adjacent to the Klamaths and then remember that this is still Eocene time and the seafloor on which it is built is moving landward as though it were a ponderously slow conveyor belt. The effect will be to telescope the rocks in that thick coastal accumulation, jamming the ones deposited offshore under those deposited nearshore and crumpling the whole pile into tight folds. The Eocene rocks in the southern end of the Coast Range are tightly folded simply because they were colliding with the Klamaths while those farther north were still riding peacefully along on their part of the seafloor.

The Willamette Valley probably rose above sea level along with the Coast Range and indeed there was probably nothing to distinguish it from the Coast Range at that early stage in its development. The Willamette Valley was dry land by the mid-dle of Miocene time, about 20 million or so years ago. Some of the big flood-basalt flows erupted from the Grande Ronde vol-cano in the northeastern corner of Oregon made it all the way to the mouth of the Columbia River to cover parts of the north end of the Coast Range and the northern part of the Willamette Valley. We can be quite sure that the Willamette Valley was above sea level then because those flows are solid, not pillow basalts as they would have been had they poured into water. The same flows did reach the ocean and become pillow basalts in the north end of the Coast Range.

It seems far more likely that the Willamette Valley did not become a lowland until long after the flood basalts had covered its northern part. In fact, the Willamette Valley is not a simple lowland at all but a series of broad basins filled with gravelly sediment and separated by tracts of low hills eroded into the basalt flows. The easiest way to imagine these basins forming is to relate them to the northward movement of the west coast during the past 15 million or so years.

There is no doubt that the Pacific plate is indeed moving northward, tearing the western edge of our continent into big slices and dragging them along. This movement is most boldly expressed in California where the famous San Andreas fault is carrying a western slice of the state northward at an average rate of about 2 inches per year. That is fast enough to move something 300 miles in a little less than 10 million years and the slice of California west of the San Andreas fault seems to have moved approximately 350 miles since Miocene time. The situation in Oregon is considerably more complicated, making the rate of movement and total displacement much less and its geologic expression quite different.

Just a few hundred miles offshore from the Pacific northwest there is an active rift in the ocean floor where two seafloor plates are pulling away from each other. The rift trends towards the northeast and, as it happens, the seafloor is pulling away from it and heading toward our coast, moving towards the southeast, away from the rift. If this were the whole story, the coast of Oregon would be carried southeastward. But the entire ocean floor, rifts and all, is moving northward and the two directions of movement tend to cancel each other. Evidently, the rate of northward movement of the entire ocean floor must be somewhat greater than the rate of southeastward movement away from the rift because the western part of Oregon does seem to be moving very slowly northward.

We can't be sure just how far north the western part of Oregon has moved but several observations suggest that the distance may be fairly close to 50 miles. In northern California the southern edge of the Klamaths is about 60 miles west and 50 miles north of the northern edge of the Sierra Nevada and there is every reason to believe that the two ranges were originally one. The 60-mile offset to the west is at least 100 million years old and obviously has nothing to do with our present problem. But the 50 mile separation to the north may well be much more recent. In Washington we find the northern end of the Coast Range slab curled into a tight fold in the Olympic Mountains just south of where it jammed into the much thicker crust of Vancouver Island. If that fold were straightened out, it would make the Coast Range tens of miles longer. And in Oregon we find the Columbia River abruptly detouring 50 miles north at Portland before resuming its westerly path to the ocean. If the entire Coast Range slab could be slid 50 miles southward, the fold in the Olympic Mountains would straighten out, the Columbia River would flow straight into the ocean, and the southern edge of the Klamaths would be directly west of the northern edge of the Sierra Nevada. Everything would fit back together into a much simpler pattern. These observations certainly don't prove that the Coast Range has moved 50 miles northward; because such things are hardly ever proveable in any final sense, but they do make it seem plausible.

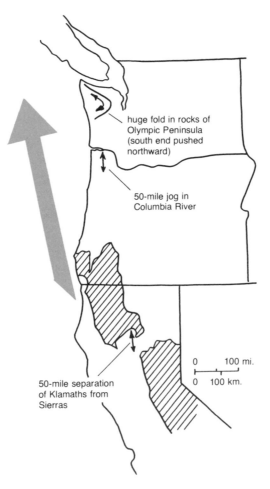

Map of the western Pacific Northwest showing relations that suggest northward movement of the coast.

It does seem reasonable that the Coast Range might move northward as a reasonably intact slab because it is close enough to the descending slab of seafloor beneath it that the two are probably fairly closely coupled together. But as the descending slab moves east it also sinks deeper so its linkage to the Oregon plate must become looser eastward. We can imagine the Coast Range moving northward, leaving the area east of it behind with a zone of tearing between them. The Willamette Valley appears to encompass that zone of tearing.

Usually a shearing motion between two parts of the earth's crust is expressed in long, straight faults which move horizontally; the San Andreas fault in California is a good example. But sometimes such shearing motion is distributed through a complex pattern of numerous smaller faults some moving horizontally and others vertically. The shearing motion between the northward moving Coast Range and the relatively stationary area east of it seems to be distributed through the Willamette Valley in a complex pattern of faults resembling that in southeastern Oregon.

It is difficult to be as confident about those faults because very few have been recognized in the Willamette Valley where they aren't as obvious as in the desert. Faults are hard to find in the rather nondescript lava flows and loose gravelly sediments that floor the Willamette Valley and the fact that they are all covered by deep soils and dense vegetation makes a difficult situation nearly impossible. But enough faults are known to faintly suggest a pattern similar to that in southeastern Oregon and the distribution and outlines of the areas underlain by young basalt and basin-fill deposits also suggest such a pattern.

The southern end of the Willamette Valley is defined by a northwest-trending fault that horizontally offsets the Cascades and dictates the straight upper course of the Willamette River southeast of Eugene. South of the Eugene area, until it meets the Klamaths near Roseburg, the Coast Range extends eastward to the Cascades just as it must have done farther north before faulting opened the Willamette Valley. This arrangement probably reflects the differences in thickness of the earth's crust north and south of Eugene.

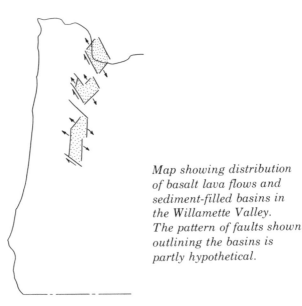

Map showing distribution of basalt lava flows and sediment-filled basins in the Willamette Valley. The pattern of faults shown outlining the basins is partly hypothetical.

The kinds of lavas erupted in the Cascades suggest that the buried northern tip of the Klamaths must extend about as far as Eugene. North of Eugene the Cascades stand on the same rocks we see in the Coast Range, on the stagnant slab of seafloor that stopped moving about 35 million years ago. The earth's crust under the Klamaths and indeed most of southern Oregon must be at least 20 miles thick, while that under the old seafloor slab is much thinner. Therefore the linkage of the descending slab of Pacific Ocean floor to the crust under southern Oregon must be much tighter and extend much farther inland than that to the thinner seafloor crust beneath northwestern Oregon. This may well explain why the shear between the sinking slab and the continent is distributed all the way across southern Oregon but is confined to the narrow zone of the Willamette Valley in the northern part of the state.

Almost all the gravelly sediments that fill the basins in the Willamette Valley washed into them during Pliocene time, between 11 and 3 million years ago. That was a period of widespread gravel deposition not only in the Willamette Valley but elsewhere in Oregon as well as in the great valley of

California, the mountain valleys of the Rockies and the length and breadth of the high plains. Had the geologic periods been named in the American west, instead of in western Europe, the Pliocene might well have been called the "Gravel period." Ever since the end of Pliocene time, our modern streams have been busily carrying all that gravel to the ocean but their job is hardly begun.

If we look at our modern world, which can't be all that different from the world of 10 million years ago, we find that widespread deposition of gravel happens only in very arid regions. Deserts have a very high rate of soil erosion because of their scanty plant cover and a very low volume of stream flow to carry the eroded debris away. So gravels eroded from desert hillsides tend to wash into the nearest valley and stay there. Evidently western Oregon was a desert during Pliocene time, incredible as that might seem today, and the basins in the Willamette Valley filled with gravel as quickly as fault movements opened them. In fact, they filled to overflowing, covering the bounding faults with gravel making them hard to recognize and the geologic picture hard to interpret today.

Even though those faults are largely buried under the Pliocene gravels and are therefore to some extent conjectural, there is nothing speculative about the earthquakes they occasionally cause. There aren't many earthquakes in the Willamette Valley, no doubt because the rate of movement is very slow, but there are a few and they do fit into the general geologic pattern. The Portland earthquake of 1962, for example, was caused by horizontal movement on a northwest-trending fault on which the south side moved northwest. This movement precisely fits the pattern we suggest and almost certainly indicates that the Coast Range is still going north and the Willamette Valley still widening.

If this movement continues for millions of years into the future, as it may well do, it will almost certainly eventually convert the Willamette Valley into the narrow inland seaway that it never has been.

interstate 5

portland — eugene

The interstate highway runs the length of the Willamette Valley between Portland and Eugene. There are very few bedrock exposures anywhere in the valley so the geology is expressed mainly in the scenery.

Portland is in a fault-block basin which, like the others farther south, is filled with sand and gravel washed into it during Pliocene time. That was between 10 and 3 million years ago when the climate of western Oregon must have been very dry; probably like the modern climate of southeastern Oregon where fault-block basins are filling with sediment today. The hills both east and west of Portland are clusters of small Cascade volcanoes all active within the last few million years. The volcanic hills west of Portland perch on a foundation of basalt lava flows that spilled down the Columbia River from northeastern Oregon.

The largest meteorite ever discovered in the United States, the Willamette meteorite, turned up in 1902 in a patch of woods that has since become West Linn. It weighs more than 15 tons. The man who found it spent several months laboriously winching it out of the woods, 100 feet at a haul, onto his own property with the idea of charging visitors admission to see it. He succeeded instead in being sued by the owners of the property from which he had hauled the stone. One law suit led to another, and then to still another, and the matter finally went to the state supreme court which awarded the meteorite to the original property owners. They eventually sold it to a lady in New York for $26,000 and she gave it to the American Museum of Natural History where it still remains, in the Hayden Planetarium. The University of Oregon has a fragment that weighs less than a half pound.

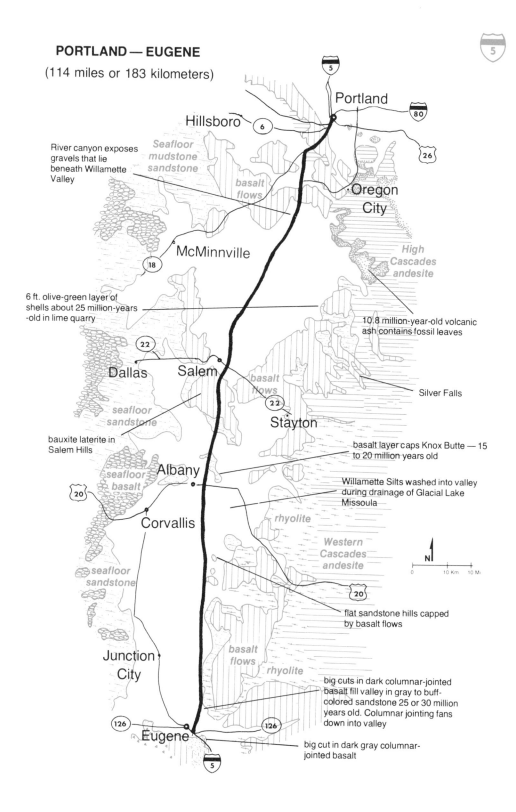

PORTLAND — EUGENE

(114 miles or 183 kilometers)

Portland

Hillsboro 6

Seafloor mudstone sandstone

River canyon exposes gravels that lie beneath Willamette Valley

basalt flows

Oregon City

High Cascades andesite

McMinnville

18

6 ft. olive-green layer of shells about 25 million-years -old in lime quarry

10.8 million-year-old volcanic ash contains fossil leaves

22

Dallas   Salem

basalt flows

22

Silver Falls

Stayton

seafloor sandstone

bauxite laterite in Salem Hills

basalt layer caps Knox Butte — 15 to 20 million years old

seafloor basalt

Albany

Willamette Silts washed into valley during drainage of Glacial Lake Missoula

20

Corvallis

rhyolite

Western Cascades andesite

N

0   10 Km   10 Mi

20

seafloor sandstone

flat sandstone hills capped by basalt flows

Junction City

basalt flows

rhyolite

big cuts in dark columnar-jointed basalt fill valley in gray to buff-colored sandstone 25 or 30 million years old. Columnar jointing fans down into valley

126

Eugene   126

big cut in dark gray columnar-jointed basalt

5

5

80

26

The interstate highway crosses low hills eroded in basalt lava flows from the Columbia Plateau in the area between Oswego and the Willamette River bridge. These hills separate the fault-block valley that contains Portland from the much larger one that the highway crosses on its way to Salem.

In the area about 10 miles south of Salem, the highway passes through the Salem hills, another tract of rolling country also eroded into plateau basalt lava flows that spilled down the Columbia River during the big eruptions that flooded much of Oregon about 20 million years ago.

Like the other such lava flows in the Willamette Valley and northern Coast Range, those in the Salem hills are capped with a thick crust of red laterite soil some of which is actually bauxitic. Laterites are soils that form in wet tropical climates and bauxite is an extreme form which is so rich in aluminum that it is used for ore. They cover the lava flows in the Willamette Valley but are themselves covered in places by the sand and gravel deposits that fill the fault-block basins and lap onto the flanks of the hills. Evidently these soils developed after the lava flows erupted but before the basins formed and filled with sediment. That means they must have formed during Miocene time and they tell us that the climate of western Oregon must have been very wet and warm then.

Modern bauxitic laterites form only in regions with a really hot climate, most of them within the truly equatorial part of the tropics. So it seems that Oregon's climate must have been hot during Miocene time when those soils developed. The fact that Miocene bauxitic

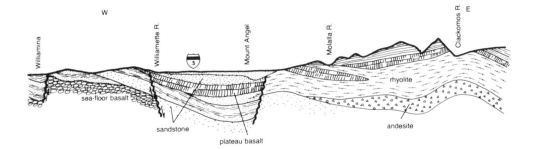

Section across the line of Interstate 5 between Portland and Eugene.

laterites exist at high latitudes in many parts of the world, not just in Oregon, suggests that for some unknown reason the whole world may have been hotter then than it is now. Perhaps the sun shone brighter during Miocene time.

South of the Salem hills, near Turner, the highway enters another broad fault-block basin which it crosses all the way to Eugene. This basin, like those to the north, is floored by thick deposits of sand and gravel washed into it during Pliocene time when the climate was very dry. These are covered in places by a skin of fine silt which washed into the Willamette Valley sometime during the last ice age when Glacial Lake Missoula drained. That was one of the strangest episodes in the geologic history of the northwest.

Glacial Lake Missoula filled many of the big mountain valleys of western Montana after a large glacier came down from British Columbia into the Idaho panhandle where it dammed the Clark Fork River. At its maximum filling the lake contained about 500 cubic miles of water. But it didn't stay filled long because the ice dam soon broke suddenly releasing all that water into the Columbia River in the biggest known flood disaster. The wall of water rushed across the Columbia Plateau, down the Columbia River, scouring its valley all the way, and then seems to have piled up behind some kind of temporary ice dam where the river jogs north at Portland. Floodwaters backed up in the Willamette Valley to an elevation of about 400 feet, laying down deposits of fine silt which contain occasional boulders that could only have come from western Montana.

Small hills capped by basalt flows dot the floor of the Willamette Valley east of the highway a few miles south of the Albany area.

interstate 5

eugene — roseburg

The Willamette Valley ends at a fault at Eugene. Between there and Roseburg the interstate highway crosses the westernmost fringe of the western Cascades and a part of the Coast Range.

The rocks between Eugene and the area about 7 miles south of Cottage Grove are andesites erupted in the first stage of western Cascade activity between about 45 and 35 million years ago. They are part of the same volcanic chain that appears east of the Cascades in the Clarno Formation of the Ochoco Mountains.

All the rocks between the western Cascades and Roseburg belong to the Coast Range. All began on the floor of the Pacific Ocean either as basalt lava flows of the bedrock ocean floor or muddy sediments that were dumped on top of them.

This is mercury country. The Bonanza Mine, the largest mercury producer in Oregon and one of the largest in the country, operated for nearly a century in the hills about 7 miles directly east of Sutherlin. Nearly 40,000 flasks of mercury, each weighing 76 pounds, came out of that deposit before the mine finally closed in 1961. The Black Butte Mine, which produced about 16,000 flasks of mercury between 1900 and 1957, is about 17 miles south of Cottage Grove and there were smaller mines and prospects in the Elkhead area.

Mercury occurs in the mineral cinnabar which is easy to spot because it looks like splashes of brilliant red paint in the rock. Hot water circulating along faults introduces it into the rocks. The mines east of Interstate 5 between Eugene and Roseburg operated in several different kinds of rock which had little in common except that they were cut by faults and loaded with cinnabar.

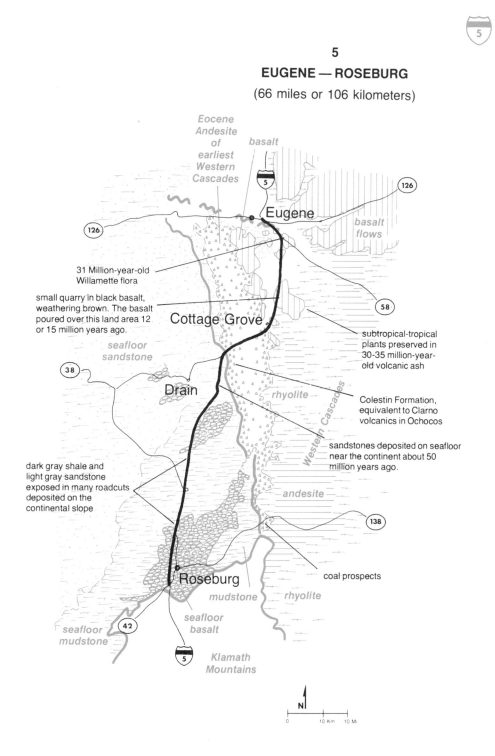

Eocene Andesite of earliest Western Cascades

basalt

126

Eugene

126

basalt flows

31 Million-year-old Willamette flora

small quarry in black basalt, weathering brown. The basalt poured over this land area 12 or 15 million years ago.

Cottage Grove

58

subtropical-tropical plants preserved in 30-35 million-year-old volcanic ash

seafloor sandstone

38

Drain

rhyolite

Colestin Formation, equivalent to Clarno volcanics in Ochocos

Western Cascades

sandstones deposited on seafloor near the continent about 50 million years ago.

dark gray shale and light gray sandstone exposed in many roadcuts deposited on the continental slope

andesite

138

Roseburg

coal prospects

mudstone

rhyolite

42

seafloor basalt

seafloor mudstone

5

Klamath Mountains

N

0   10 Km   10 Mi

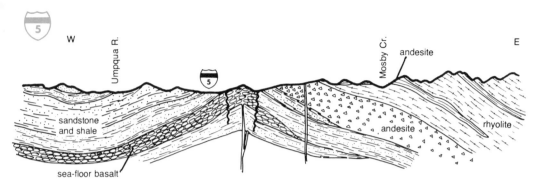

Section across I-5 about midway between Eugene and Roseburg. Andesite and rhyolite east of the highway belong to the western Cascades; they rest on Coast Range rocks.

Cinnabar is a very unstable mineral which breaks down and releases mercury vapor if it is roasted at a moderate temperature. The old mercury mines roasted their ores in rather simple ovens that usually leaked and then condensed most of the mercury vapor but always managed to lose some into the air. Nothing is more poisonous than mercury vapor which insidiously attacks and finally destroys the central nervous system with insanity being one of the more obvious symptons. The mercury mines were extremely unhealthy places to work.

Alternating thin layers of sandstone and mudstone that once lay on the floor of the Pacific Ocean are now exposed in a roadcut near Sutherlin.

u.s. 20

newport — albany

The route between Newport and Albany passes through the Coast Range to the Willamette Valley. All the rocks along the way were once part of the Pacific Ocean floor; some are basalt lava flows that formed the bedrock floor of the ocean and the rest are soft mudstones and sandstones deposited on the basalt.

Rocky outcrops are scarce along U.S. 20 as they are nearly everywhere in the densely forested Coast Range where the bedrock is thickly covered with soil. This is frustrating country for geologists. It is a thrill to walk around on an old sea floor but it would be a lot more fun if we could just see a little more bedrock.

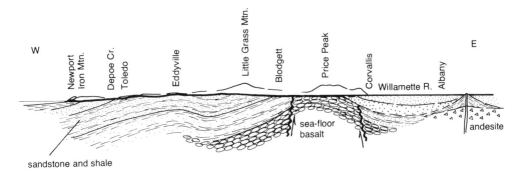

Section along the line of U.S. 20 between Newport and Albany. The entire northern part of the Coast Range is simply a big slab of sea floor raised high and dry, tilted ever so gently eastward, and broken up a bit by a few faults.

Newport is on a small area of sediments deposited during Miocene time, about 20 million years ago, long after all the Coast Range had raised above sea level except for the area around Astoria and a few small sections along the coast. About 3 miles inland from Newport the road leaves these younger sediments and gets onto the older

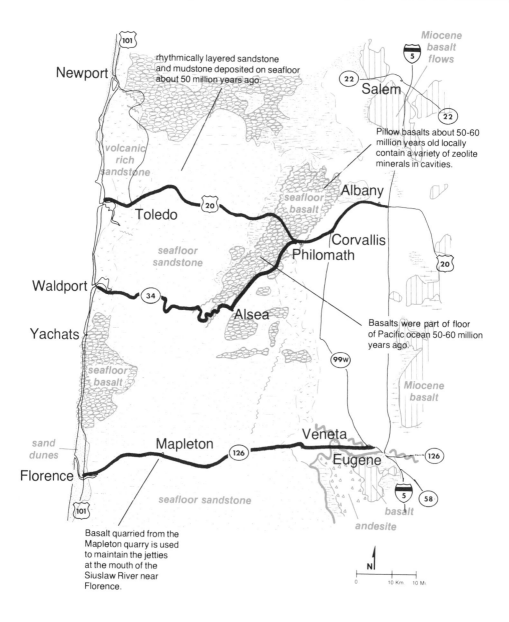

20
ALBANY — NEWPORT
(54 miles or 87 kilometers)

34
CORVALLIS — WALDPORT
(64 miles or 103 kilometers)

126
EUGENE — FLORENCE
(61 miles or 98 kilometers)

Newport

rhythmically layered sandstone and mudstone deposited on seafloor about 50 million years ago.

Miocene basalt flows

Salem

volcanic rich sandstone

Pillow basalts about 50-60 million years old locally contain a variety of zeolite minerals in cavities.

seafloor basalt

Albany

Toledo

seafloor sandstone

Corvallis

Philomath

Waldport

Yachats

Alsea

Basalts were part of floor of Pacific ocean 50-60 million years ago.

seafloor basalt

Miocene basalt

sand dunes

Veneta

Mapleton

Florence

seafloor sandstone

Eugene

basalt andesite

Basalt quarried from the Mapleton quarry is used to maintain the jetties at the mouth of the Siuslaw River near Florence.

N

0   10 Km   10 Mi

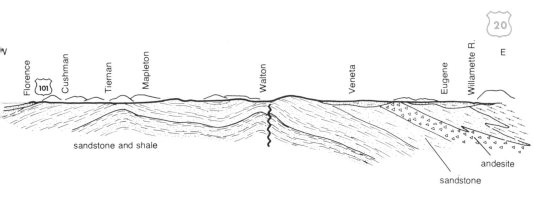

Section across the Coast Range along the line of highway 126 between Florence and Eugene.

mudstones deposited far out on the deep ocean floor while it was still moving landward.

About two miles east of Blodgett the road leaves the sedimentary rocks and crosses onto an area of basalt sea floor raised as a block by movement along faults. Just west of Philomath the sand and gravel floor of the Willamette valley laps onto the eastern flank of the Coast Range. These deposits accumulated several million years ago when western Oregon had a dry climate and the drainage wasn't powerful enough to carry such debris away to the ocean. U.S. 20 crosses the old valley-fill deposits and the modern floodplain of the Willamette River between Philomath and Albany.

Waves pound a basalt coastline at Yachats Bay.

u.s. 26

seaside — portland

The route cuts right through the Coast Range across two basically different major assemblages of rocks. The older of these is the complex mess of sediments and basalt lava flows formed on the floor of the Pacific Ocean and the younger consists simply of large basalt lava flows, some of which may be the westernmost end of the Columbia lava plateau. As everywhere in the Coast Range, vegetation is dense and the soil cover thick so the bedrock is poorly exposed and the geology difficult to appreciate either along the road or anywhere else.

All the rocks between Seaside and Buxton are lava flows and sediments formerly on the seafloor. They are youngest near the coast and get progressively older inland. Rocks between Seaside and Necanicum Junction are mostly dirty sandstones and mudstones deposited along the continental margin during Miocene time, between 15 and 20 million years ago. They contain a number of large basalt lava flows, one of which is nicely exposed in roadcuts immediately east of Necanicum Junction.

These flows evidently erupted from a nearby chain of Miocene volcanoes aligned in a north-to-south trend approximately following the present coastline. Several of those volcanoes now form prominent

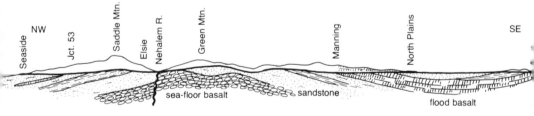

Section along the line of U.S. 26 between Seaside and Portland.

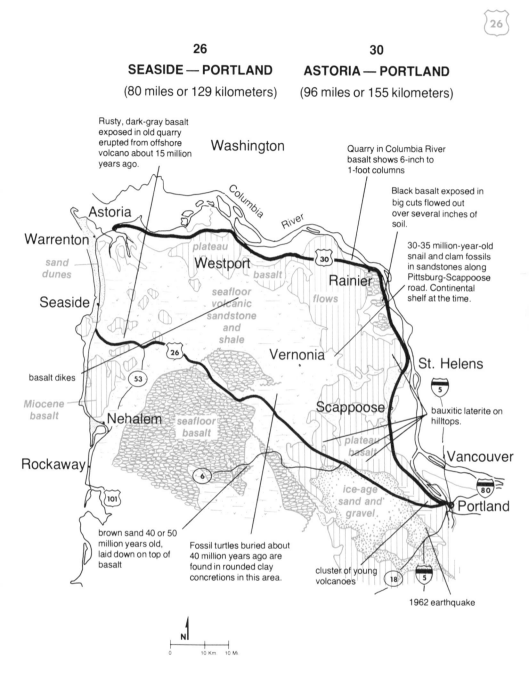

26

SEASIDE — PORTLAND

(80 miles or 129 kilometers)

30

ASTORIA — PORTLAND

(96 miles or 155 kilometers)

Rusty, dark-gray basalt exposed in old quarry erupted from offshore volcano about 15 million years ago.

Washington

Quarry in Columbia River basalt shows 6-inch to 1-foot columns

Columbia River

Black basalt exposed in big cuts flowed out over several inches of soil.

Astoria

Warrenton

sand dunes

plateau

Westport

basalt

30

Rainier

30-35 million-year-old snail and clam fossils in sandstones along Pittsburg-Scappoose road. Continental shelf at the time.

Seaside

seafloor volcanic sandstone and shale

flows

basalt dikes

26

53

Vernonia

St. Helens

5

Miocene basalt

Nehalem

seafloor basalt

Scappoose

bauxitic laterite on hilltops.

Rockaway

6

plateau basalt

Vancouver

80

101

ice-age sand and gravel

Portland

brown sand 40 or 50 million years old, laid down on top of basalt

Fossil turtles buried about 40 million years ago are found in rounded clay concretions in this area.

cluster of young volcanoes

18

5

1962 earthquake

N

0   10 Km   10 Mi.

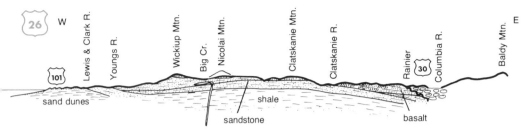

Section south of the line of U.S. 30 between Astoria and Rainier.

capes on the northern Oregon coast. Two others, Saddle Mountain and Humbug Mountain, are outstanding landmarks a few miles north of U.S. 26 about halfway between Necanicum Junction and Elsie. There is a state park at Saddle Mountain which contains beautiful exposures of pillow basalts; apparently these flows erupted underwater.

Between Necanicum Junction and Elsie, the road crosses older dirty sandstones and shales deposited on the seafloor about 40 million years ago. Most contain volcanic ash derived from nearby volcanoes.

The hills south of the highway between Elsie and Buxton are made almost entirely of pillow basalts which appear to be the 55-million-year-old bedrock floor of the ocean, not local lava flows erupted from a chain of younger volcanoes like those farther west. Very little of the basalt is readily visible because the road is routed through the gentler topography eroded on the softer sedimentary rocks. But the road builders paid for their easy route in another way because the soft sediments are very weak and prone to landsliding. There is an especially large area of landsliding about 4 miles east of Elsie.

Between Buxton and Portland the bedrock is basalt lava flows that seem to be the western end of the Columbia Plateau. They are the same age as the flows near the coast but show no sign of having erupted underwater and are similar in every way to the plateau flows farther east.

Oregon had a wet, tropical climate during Miocene time when the plateau flows erupted and they are covered by deep, red soils, called laterites, which typically develop under those conditions. Red Miocene soils are noticeable in road cuts a short distance east of Buxton and here and there on into Portland.

Weak bedrock, deep soils, and a wet climate make landsliding a constant problem in the Coast Range. This one threatens U.S. 26 a few miles west of Buxton.

Quite a number of bauxite deposits exist in the general area between Buxton and Portland. Bauxite is a kind of lateritic soil that is very rich in aluminum oxide and poor in iron oxide. Unfortunately, none of the Oregon bauxites are quite rich enough in aluminum oxide or poor enough in iron oxide to compete with imported ores although some come close and may someday be mineable.

All of the Oregon bauxites are developed on basalts erupted during Miocene time, between 25 and 15 million years ago, and almost all of them cap broad hills and ridges — the reason for that is not clearly understood but it seems to have something to do with circulation of ground water. We know that Oregon had a wet, tropical climate back in Miocene time and it seems likely that the bauxite deposits, indeed most of the Oregon laterites, developed then.

Immediately west of Portland, on the western flank of the Portland hills, there is a small chain of volcanoes the largest of which is Mount Sylvania. These volcanoes were active during late Pliocene time, perhaps 3 to 5 million years ago. Lava flows erupted from them bury silts that bury the bauxites a few miles farther west. So the silts must be fairly old and this seems to be good evidence that the bauxites probably did develop during Miocene time.

u.s. 30

astoria — portland

The road follows close by the Columbia River as it cuts its sweeping northward curve through the Coast Range on its way from Portland to the sea. Sedimentary rocks form the bedrock along part of the route and flood basalt lava flows from the big Columbia Plateau eruptions in northeastern Oregon along the rest.

Astoria is built on mudstones laid down on shallow seafloor during Miocene time when most of the Oregon Coast Range was already above sea level. They are the same age as the big basalt flows that make the Columbia Plateau and actually poured out into the sea in this part of the Coast Range. The Astoria mudstones contain fossils of animals much more tropical in their affinities than any that lived along the Oregon Coast either before or after Miocene time, the period between about 11 and 25 million years ago. The fossils suggest that the climate was much warmer then and the tropical laterite soils on top of the lava flows repeat the message.

Along much of the route between the eastern outskirts of Astoria and the Bradwood area, the bedrock is sandstone which appears to have deposited on land during Pliocene time when the climate was arid. The sandstones along U.S. 30 appear to have deposited in an old valley of the Columbia River which must have almost dried up during Pliocene time.

Between the Bradwood area and Marshland the route follows the broad floodplain of the Columbia. The bluffs south of this stretch of highway are capped by the flood basalt lava flows that erupted in northeastern Oregon and poured all the way into this area during Miocene time while the north end of the Oregon Coast Range was still partly submerged. Between Marshland and Rainier the road crosses low hills eroded into those same flows and then passes bluffs capped with them as it follows the river into Portland.

The older rocks beneath the flood basalt flows are well exposed in places along the highway between Rainier and Goble. Most of them are also basalt lava flows, but much older ones, erupted on the seafloor nearly 50 million years ago. Like all sea floor basalts, they look like piles of big, dark pillows.

Some of the flatter hilltops on the flood basalt flows are capped with bauxitic laterite, a tropical soil that develops under very warm and humid climatic conditions. They are the same age as the tropical fossils in the Astoria mudstones so the pieces of the puzzle fit together rather well.

The best bauxite deposits are full of gibbsite, a white mineral that looks almost like paraffin wax and is nearly pure aluminum oxide. The gibbsite forms round lumps which may be as small as peas or as large as potatoes, they show up in numerous cuts along secondary roads back in the hills and litter the fields in some hilltop farms in this area.

All of the Oregon bauxites are red because they contain some iron oxide. None are as pure as the high-grade ores imported from the Caribbean region which are almost white. Iron oxide must be separated from the ore before it is reduced to the metal and that extra processing step raises the cost of production higher than it does to import ore. But there is a lot of bauxite in Oregon which may be mined some day if a better method of separating iron oxide is developed or the foreign supply is cut off. Since the bauxite is essentially just a soil, it would be mined from the surface in open pits.

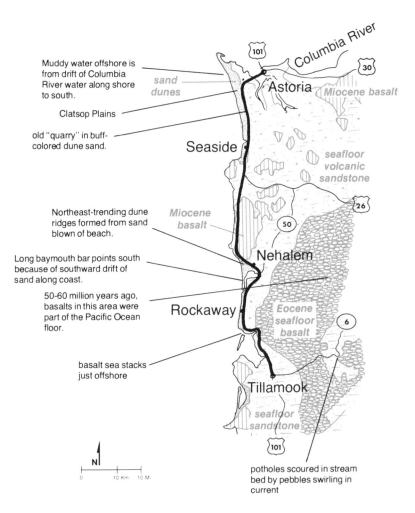

Muddy water offshore is from drift of Columbia River water along shore to south.

sand dunes

Astoria

Miocene basalt

Clatsop Plains

old "quarry" in buff-colored dune sand.

Seaside

seafloor volcanic sandstone

Northeast-trending dune ridges formed from sand blown of beach.

Miocene basalt

Long baymouth bar points south because of southward drift of sand along coast.

Nehalem

50-60 million years ago, basalts in this area were part of the Pacific Ocean floor.

Rockaway

Eocene seafloor basalt

basalt sea stacks just offshore

Tillamook

seafloor sandstone

Columbia River

N

0   10 Km   10 Mi

potholes scoured in stream bed by pebbles swirling in current

The remains of the Peter Iredale *moldering into rust in Fort Stevens State Park at the mouth of the Columbia River.*

u.s. 101

astoria — tillamook

Astoria sprawls over low hills eroded into soft mudstones deposited during Miocene time, roughly 20 million years ago, when this part of the Coast Range was still under water. Occasional higher hills in the vicinity are underlain by basalt lava flows erupted offshore during the same period.

The Astoria mudstones are dark gray where they are freshly exposed, but a bit of weathering quickly produces iron oxides which stain them reddish orange, the color most often seen except in new excavations. They contain a few big fossils and quite a variety of microscopic ones, all remains of animals that would have been comfortable living in water much warmer than that along the modern coast.

The Astoria mudstones are very weak rocks which weather rather easily to form deep soils which have approximately the consistency and mechanical strength of grease when they get thoroughly soaked. Landslides have been a problem along this part of the coast for years and will certainly get worse unless future development on the steeper slopes is planned to avoid making them any steeper or wetter than they already are. Cut-and-fill terracing and septic tank sewerage systems are very treacherous things in any area underlain by such rocks. The best insurance against landslides is to avoid undercutting slopes and to maintain good drainage.

Between the outskirts of Astoria and the north side of Tillamook Head, U.S. 101 runs the length of the giant Clatsop sand spit. Waves and wind working the sediment brought to the coast by the Columbia River have built most of this vast, sandy plain since sea level rose to its present stand about 8500 years ago after the last ice age ended. During the winter months the prevailing winds blow from the southwest driving sand brought in during that season northward along a corrresponding beach in Washington. Neither beach will grow as fast in the future as in the past because the series of dams along the Columbia now traps most of the sediment before it reaches the coast.

This dense stand of European beach grass created the beach ridge on which it grows by trapping sand blown inland by the wind. The imported grass is transforming the Oregon beachscape.

Onshore winds play their part by sweeping sand off the drying upper beach while the tide is out and blowing it inland into the dunes. Clatsop sand spit had spectacular tracts of constantly shifting sand dunes until the 1930's when most of them were deliberately killed by artificial plantings of grass and shrubs which included species not native to the Oregon coast. Many lovely tracts of sand dunes along this coast have met the same sad fate, most often for no apparent reason except that they were moving as nature intended. Most of the dunes that used to march rakishly across the Clatsop spit threatened nothing. Shapeless heaps of stabilized sand huddling beneath a cover of planted vegetation are neither as beautiful nor as interesting as living dunes sculptured by the wind and responding to every shift in the mood of the season.

Wind sculpture in sand.

Big sand spits are not really as level as they look from a distance. They always contain old beach ridges that run generally parallel to the coast and are separated by low swales that often hold long ponds or peat bogs. Some of the old beach ridges on the Clatsop spit are so straight and regular that they could be mistaken for abandoned railroad embankments. They mark old positions of the shoreline left inland as the influx of new sand built the beach seaward. Tracing old beach ridges on a detailed map or aerial photograph is a good way to unravel the history of a sand spit by reconstructing its appearance at past stages of development.

Beach ridges probably start as submerged breaker bars that develop offshore where the long waves crash during heavy storms. A breaker bar is really an incipient beach and eventually the waves sweeping sand down the coast will build it up to the point where it emerges above sea level and becomes a real beach. Meanwhile, a new breaker bar forms offshore and the process repeats.

The Clatsop sand spit ends about 3 miles south of Seaside, at the mouth of the Necanicum River, and between there and Nehalem the highway winds along the coastline, crossing bedrock consisting of mudstones and basalt lava flows formed during Miocene time, about 20 million years ago. These are essentially similar to the rocks around Astoria. This combination of rocks makes a beautiful coastline full of scenic variety. Basalt is very hard and sturdily withstands the battering surf to form headlands that stand boldly out to sea but the soft mudstones yield easily to the waves, forming coves and long stretches of smooth coastline fringed by sandy beaches.

Tillamook Head exists because there is a large intrusion of basalt sandwiched between layers of mudstone which just happens to be exposed right at sea level where it can withstand the pounding surf. Had the layer of basalt been either higher or lower, Tillamook Head would not be there.

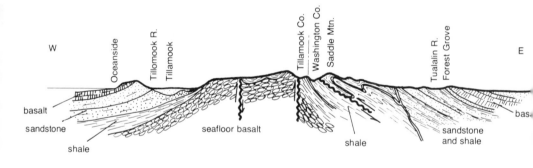

Section across U.S. 101 at Tillamook Head.

Ecola State Park at Tillamook Head is a good place to see active landsliding in the Astoria mudstones. The parking area replaces another that was broken up by slide movement during the winter of 1961. Scraps of the old pavement still stand at odd angles in the shrubbery. Scarps high on the hillslopes with patches of hummocky topography below them on which the trees tilt backward into the slope, mark several other active landslides in the park.

Between Tillamook Head and Arch Cape the bedrock is almost entirely soft mudstones in which the waves have shaped a long, smooth stretch of coast. A series of prominent mountain peaks with irregular profiles dominates the skyline of the Coast Range east of the highway. Most of these appear to be remains of volcanoes that were active during Miocene time while the mudstones in this area were accumulating on the seafloor. Most of the lava flows in those mountains are pillow basalts so obviously they erupted under water and it seems likely those volcanoes were mostly, if not entirely, submerged. Few of them broke the surface to become islands.

Hug Point, Arch Cape, and Cape Falcon all exist because large intrusions of basalt in the mudstone buttress parts of the coast against wave attack. These intrusions formed where masses of molten basalt magma squirted into the mudstones and cooled within them instead of erupting as lava flows on the seafloor. They are the same age as the numerous Miocene volcanoes and lava flows in this part of the Coast Range and certainly closely related to them. Observation of the contact of the intrusions shows that the molten basalt squirted into soft, sloppy mud — not solid mudstone.

Neahkanie Mountain is a pile of basalt erupted during Miocene time, about 20 million years ago, when this part of the Coast Range was still underwater. It was probably a volcanic island then standing a few miles offshore. From its south face to Tillamook the waves break on a coastline made of soft mudstones and sandstones which they erode very easily. Wide, sandy beaches fringe this stretch of coast as they do wherever the surf attacks soft sandstones.

u.s. 101

tillamook — florence

Except for the northernmost stretch between Tillamook and Oretown, this entire route clings tightly to the coast so the ocean and the beach are almost constantly in view. Bedrock is continuously exposed in the sea cliffs and in many places along the road. Most of it is sediments and pillow basalts but younger volcanic rocks buttress some of the headlands against wave attack.

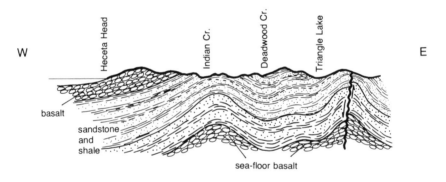

Section across U.S. 101 between Yachats and Florence.

Between Tillamook and Oretown, U.S. 101 follows an inland route across sandstones, mudstones, and basalts that were formerly on the floor of the Pacific Ocean. Secondary roads venture out to Capes Meares and Lookout; both spectacular headlands composed of basalts erupted during Miocene time, about 15 or 20 million years ago.

The map view of Cape Lookout suggests a dagger stabbing toward the sea. Its wave-scrubbed cliffs expose a wonderful sequence of lava flows, some of them the solid kind that cooled on dry land and others pillow basalts that obviously erupted under water. Evidently the waves breaking on this pointed cape are patiently pounding the last remnants of a Miocene volcanic island into sand and mud.

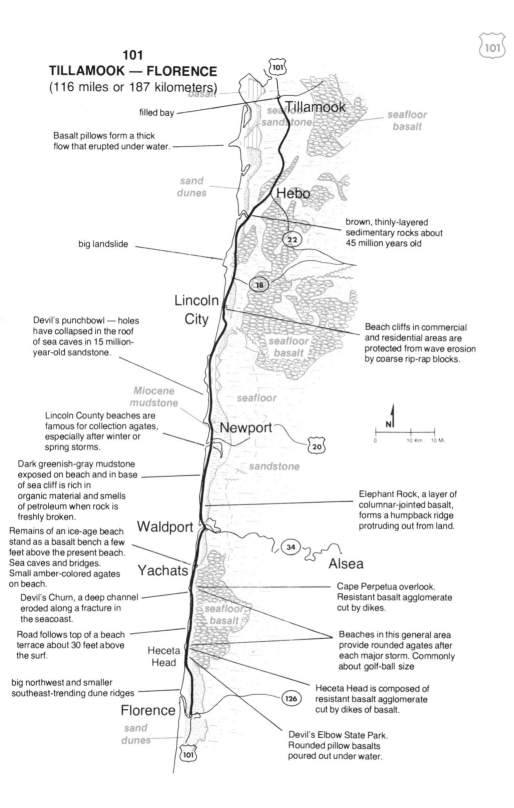

101
TILLAMOOK — FLORENCE
(116 miles or 187 kilometers)

basalt

filled bay

Tillamook

seafloor sandstone

seafloor basalt

Basalt pillows form a thick flow that erupted under water.

sand dunes

Hebo

brown, thinly-layered sedimentary rocks about 45 million years old

big landslide

Lincoln City

Devil's punchbowl — holes have collapsed in the roof of sea caves in 15 million-year-old sandstone.

seafloor basalt

Beach cliffs in commercial and residential areas are protected from wave erosion by coarse rip-rap blocks.

Miocene mudstone

seafloor

Lincoln County beaches are famous for collection agates, especially after winter or spring storms.

Newport

Dark greenish-gray mudstone exposed on beach and in base of sea cliff is rich in organic material and smells of petroleum when rock is freshly broken.

sandstone

Elephant Rock, a layer of columnar-jointed basalt, forms a humpback ridge protruding out from land.

Waldport

Remains of an ice-age beach stand as a basalt bench a few feet above the present beach. Sea caves and bridges. Small amber-colored agates on beach.

Yachats

Alsea

Devil's Churn, a deep channel eroded along a fracture in the seacoast.

seafloor basalt

Cape Perpetua overlook. Resistant basalt agglomerate cut by dikes.

Road follows top of a beach terrace about 30 feet above the surf.

Heceta Head

Beaches in this general area provide rounded agates after each major storm. Commonly about golf-ball size

big northwest and smaller southeast-trending dune ridges

Florence

Heceta Head is composed of resistant basalt agglomerate cut by dikes of basalt.

sand dunes

Devil's Elbow State Park. Rounded pillow basalts poured out under water.

N

0   10 Km   10 Mi

Cape Kiwanda, just a few miles north of Oretown, is a small but extraordinarily beautiful headland made of sandstone intricately sculptured by the waves. Sandstones in this part of the Coast Range are too soft to stand against the surf and Cape Kiwanda would not exist were it not for a large sea stack made of basalt that stands offshore just beyond its tip absorbing the impacts of the waves.

Bedrock exposed in the sea cliffs between Oretown and Lincoln Beach is sandstone, mudstone and basalt formed on the seafloor during Eocene time, 50 million or so years ago. The softer sedimentary rocks outcrop along the smooth stretches of coastline, whereas the harder basalts are exposed in the rocky coast between Neskowin and Cascade Head. Those same basalts extend inland almost as far as the Willamette Valley, making a large exposure of the old bedrock seafloor.

Cape Foulweather is a complex of volcanic rocks; those near its tip appear to have formed near a volcanic vent and those to the north and south are the kinds of rocks normally seen on the flanks of a volcano. There are solid lava flows that erupted on dry land and pillow basalts that erupted under water. And there are beds of volcanic ash, especially well exposed around Government Point, some of which appear to have settled directly from the air and others in currents of water. Cape Foulweather is obviously the wreck of a volcanic island that stood a short distance offshore during Miocene time and has since raised above sea level as the Coast Range rose.

Bedrock between Cape Foulweather and the area just north of Seal Rock is sandstone and mudstone deposited offshore during Miocene time, about 20 million or so years ago, while the Cape Foulweather volcano was active. This short stretch of coast, like the area around Astoria, remained submerged long after the rest of the Coast Range had risen above sea level.

Yaquina Head is the last remnant of another Miocene volcano now standing as a stalwart outpost of basalt jutting sharply into the surf. But its resistance will eventually be futile because no rock is sturdy enough to stand indefinitely against the untiring surf. The softer rocks yield first, leaving the harder ones standing isolated as headlands. Then the waves focus their energy on the headland as they curl around to batter it from both sides and send huge breakers crashing against its tip. Such concentrated fury is bound to prevail and the headland will erode away as the waves fill nearby coves with the debris of its destruction. So no matter what the original outline of a coast may be, the waves will reduce it to a series of smoothly sweeping curves as they destroy the headlands and fill the coves.

Most freshly erupted volcanic rocks are full of holes which later fill with mineral matter brought into them by circulating ground water. Sometimes the mineral fillings are agates, more often they are zeolites — a family of minerals that comes in a variety of attractive pale colors. Most old volcanic rocks are full of agates or zeolites, or both, and where waves erode them the hard ones survive to become beach pebbles.

Agates are common in many places along the Oregon coast, not just at Agate Beach. The best time to look for them is during late winter or early spring after the heavy storms have moved the sand offshore and left a pebbly beach. The gentler waves of summer bring the sand back onshore and bury most of the pebbles including the agates.

Seal Rock is another isolated mass of hard basaltic rock holding up a headland while the softer sedimentary rocks on either side of it erode more quickly to form a smooth coastline fringed by a sandy beach. The soft sedimentary rocks and sandy beach extend southward past the mouth of the Alsea River almost to Yachats. The Alsea sandspit at the north side of the river mouth has been growing rapidly southward in recent decades extending itself by as much as 10 feet each year.

Between the area about a mile north of Yachats and that just south of Heceta Head, the waves batter another stretch of hard basalt making a rocky coastline full of picturesque headlands and coves. The basalt is part of the old bedrock seafloor, lava flows erupted under water during Eocene time, 50 million or so years ago. It is not perfectly solid and uniform; volcanic rocks never are, but contains numerous zones of broken and fragmented rock among the solid lava flows. The waves exploit every weakness carving the broken rock away to make coves and leaving the stronger rocks standing as bold headlands and sea stacks.

Here and there along this stretch of coast are conspicuous remnants of an old beach created by the waves some hundreds of thousands of years ago and now about 20 to 30 feet above sea level. Either the land has risen or sea level has dropped; it is difficult to be sure along such an active coastline where uplift of the land is perfectly possible. The old beach makes a smooth terrace that slopes gently seaward and makes wonderful sites for homes and the highway; the entire town of Yachats is built on it. That old shoreline had sea stacks offshore just as the modern one does and some of them are still clearly recognizeable as big wave-carved rocks standing on the terrace.

About 8 miles north of Florence the basalt ends and a long stretch of smooth and sandy coast begins, once again because the waves are eroding soft sedimentary rocks which contain a lot of sand. Sutton Lake and Mercer Lake just north of Florence both exist because sand dunes migrating inland from the beach have dammed the drainage of small streams. The outlets of both lakes are by seepage through the sand.

Waves exploited fracture zones in the hard basalt to carve Sea Lion Caves near Florence.

u.s. 101

florence — port orford

The stretch of U.S. 101 between Florence and Coos Bay follows a coast piled high with sand. This is probably the most spectacular tract of coastal sand dunes in the country. Rivers carried some of the sand to the sea and waves battering the sandstones exposed in the seacliffs supplied the rest. Every time the tide comes in, waves carry new sand high onto the beach. They leave it there where it dries as the tide goes out and then the wind blows it inland into the dunes.

Prevailing winds blow onto the Oregon coast from the southwest in winter and the northwest in summer. So the waves tend to drive sand northward along the beach all winter and then back south again in summer. Along some sections of the coast the winter waves prevail and sandspits grow from south to north; in others the summer waves hold the upper hand and sandspits build southward. The mouth of the Siuslaw River at Florence, for example, is diverted about 3 miles north by growth of a sandspit whereas those of the Umpqua River at Reedsport and the Millicoma River at Coos Bay are both diverted several miles to the south. The reason for the difference is not obvious but probably has something to do with the shape of the coast — whether the particular beach happens to face more directly toward the southwest or northwest.

Coastal sand dunes also show the influence of the changing seasonal pattern of prevailing winds. Heavy winter storms bring strong onshore gales out of the southwest, shaping the major outlines of the dunes. Then the gentler winds of summer form cross ridges which are quite conspicuous by fall but disappear entirely before the next spring. So the dunes change in character with the passing of the seasons. Indeed, they change almost from day to day as the restless hills of sand shift to reflect every change in the weather. And every little shift erases yesterday's footprints and tire tracks leaving the sand surface fresh and unmarked, ready for the next explorer.

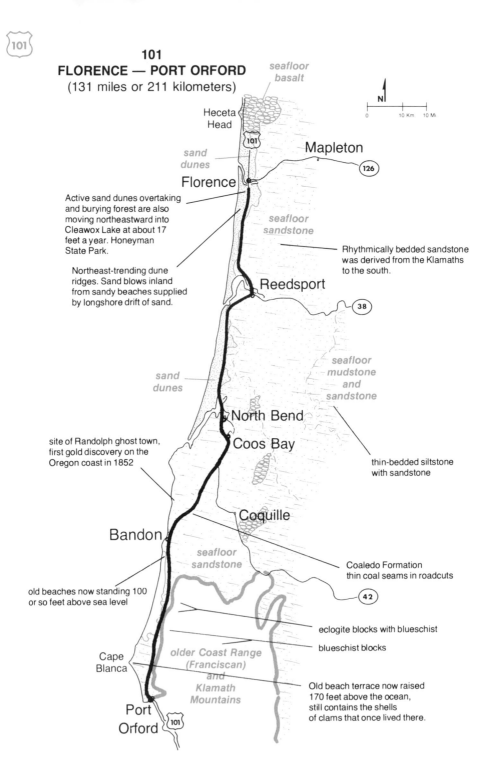

101
FLORENCE — PORT ORFORD
(131 miles or 211 kilometers)

seafloor basalt

Heceta Head

N
0   10 Km   10 Mi

Mapleton

126

sand dunes

Florence

Active sand dunes overtaking and burying forest are also moving northeastward into Cleawox Lake at about 17 feet a year. Honeyman State Park.

seafloor sandstone

Rhythmically bedded sandstone was derived from the Klamaths to the south.

Northeast-trending dune ridges. Sand blows inland from sandy beaches supplied by longshore drift of sand.

Reedsport

38

seafloor mudstone and sandstone

sand dunes

North Bend

Coos Bay

site of Randolph ghost town, first gold discovery on the Oregon coast in 1852

thin-bedded siltstone with sandstone

Coquille

Bandon

seafloor sandstone

Coaledo Formation thin coal seams in roadcuts

old beaches now standing 100 or so feet above sea level

42

eclogite blocks with blueschist

blueschist blocks

Cape Blanca

older Coast Range (Franciscan) and Klamath Mountains

Old beach terrace now raised 170 feet above the ocean, still contains the shells of clams that once lived there.

Port Orford

101

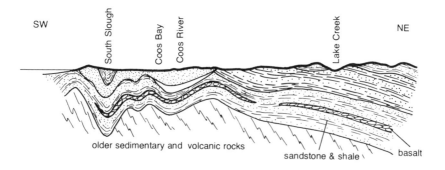

SW | South Slough | Coos Bay | Coos River | Lake Creek | NE

older sedimentary and volcanic rocks

sandstone & shale

basalt

Section across the Coos Bay area showing how the rocks there are folded down into a deep basin.

Sand grains are too heavy to stay airborne for more than a few inches so the wind can only bounce them along the surface. Blowing sand rarely rises more than a foot or so off the ground. Grains of sand bounce better on hard surfaces than on soft ones so the wind sweeps hard surfaces clean and piles the sand in soft places where it sticks. And of course a dry sand dune is very soft so any grains that bounce on to it will stick and add themselves to the dune. That is why the wind sweeps loose sand into heaping dunes instead of blowing it all over the countryside.

A brisk sea breeze sends sand smoking over the crest of a dune in Honeyman State Park near Florence.

93

Hard objects left on dunes cause severe damage by spoiling the soft surface which is the reason the dune exists. Softness is the very essence of a sand dune. Blowing sand grains bounce better when they strike a hard surface and that causes the wind to blow a hole in the dune. That is why we so often see collections of bottles, tin cans and old dune buggy parts paving the bottoms of holes in sand dunes. They aren't there because somebody threw the junk in a hole, they are there because somebody threw the junk on the dune and then the wind dug the hole. And the wind will maintain the hole as long as the junk is exposed. An easy way to fill such holes is to cover the hard junk with enough sand to restore the soft surface and then let the wind finish the job.

These shapeless hummocks west of Honeyman State Park are old dunes stabilized by European beach grass.

For some obscure reason sand dunes seem to offend many people simply because they move; hills should stay firmly rooted in place and not go wandering off. Enormous amounts of taxpayer's money are spent each year to "control" any number of perfectly harmless dunes, converting them into characterless heaps of sand. Unfortunately, these efforts at dune control almost always succeed.

The usual approach to dune control is to plant grass on the sand. The leaves break the force of the wind and also intercept flying grains, preventing them from striking the sand surface with enough force to kick others into the air. Many people suppose that roots stabilize sand dunes by somehow holding the sand but it should be obvious that they can't possibly do any such thing. Roots grow underground and wind blown sand grains fly through the air so neither has much effect on the other.

Sand blows from left to right up the slope of this dune and then slides down the steep face. The European beach grass will survive the encroaching sand by growing upward as it is buried.

None of the native grasses grow well on sand dunes so the stabilizers introduced beach grass from Europe about 1910. Oregon suits it splendidly and now European grass grows wild all along the coast, ruining many beautiful tracts of dunes and literally changing the shape of the beachscape. Sand used to blow from the beach back into the dunes but now the dense growths of grass immediately behind the beach trap most of it. So a grass-covered ridge is growing just behind the beach and the dunes are marching inland with their supply lines cut off behind them.

The numerous freshwater lakes east of the road between Florence and Coos Bay all exist because big sand dunes have buried the mouths of streams damming them. A few larger streams have enough flow to keep their channels open to the sea but smaller ones can't cope with the sand. Underground seepage through the dune provides outlets for the lakes. All of the dunes move slowly northeastward, driven primarily by the winter winds, and will slowly fill the lakes. Measurements of the rate of advance of several dunes give results in the range from 6 to 18 feet per year. It is easy to see that they must be advancing because they encroach on the forest, killing and then burying the trees.

Typical section through a sandy coastline. Dense growths of beach grass are now trapping sand in the ridge just behind the beach, making it grow, and at the same time starving the dunes.

Black sand is common along the entire length of the Oregon coast, just enough in the dunes to add an occasional dash of color but many beaches are broadly splashed with black. The sand contains a variety of minerals all denser than the quartz and feldspar that comprise most ordinary sand. Magnetite, the magnetic oxide of iron and ilmenite, a titanium mineral, are usually the most abundant but most black sands also contain red garnet, a few glittering grains of zircon, and a brown mineral called monazite that contains rare-earth elements. Some contain gold. The waves know these heavy minerals by their weight and neatly sort them into separate deposits.

Oregon black sands are more interesting than most because they generally contain chromite and platinum along with the usual minerals. A few people have made a little bit of money and a lot of others have gone broke trying to mine the beach sands. Rising prices of many mineral commodities may yet make it worthwhile.

Except for a short stretch near Bandon, almost the entire route between Coos Bay and Port Orford is well inland and away from the coast. The road passes through flat or very gently rolling countryside so densely cloaked in second-growth forest that hardly any rocks are visible.

Coos Bay formed a little less than 10,000 years ago as the last ice age ended. During the ice age sea level dropped nearly 300 feet because so much water was tied up in the big continental ice sheets

96

and the shoreline was out where the water is now that deep. Then the ice melted and sea level rose to its present stand, flooding river mouths to form estuaries such as Coos Bay. The bottom of the bay is simply the old floodplain of the Millicoma River. Waves and wind driving sand southward down the beach have built the big sandspit that forms the seaward side of outer Coos Bay since sea level rose about 10,000 years ago.

Most of the route between Coos Bay and Bandon passes through rolling hills eroded in the Coaledo Formation, a thick series of sands and muds deposited near shore, possibly in a bay, during Eocene time. This formation contains several substantial coal seams which supported large mines many years ago and surely will again. Such thick sequences of sedimentary rocks often contain oil and gas, so with luck the Coos basin may eventually produce more than coal. Wildcat drilling began back in 1919 and has continued ever since in a casual way, about a dozen wells in more than 50 years. Although several of these wells produced small quantities of natural gas and shows of oil, none were commercially worthwhile. But a dozen wells, none of them very deep, aren't nearly enough to thoroughly test the potential of an area this big so there is still some hope that deposits of oil and gas may exist. The wildcat drilling will continue. The sea cliffs around Cape Arago just west of Coos Bay are probably the best place to see good outcrops of the Coaledo Formation, especially when the tide is out.

Between the area about 6 miles north of Bandon and Port Orford, the road follows the surface of a marine terrace, that was leveled by wave action when it was submerged offshore and then raised above sea level. The old shoreline is near the base of the hills east of the highway. Big, craggy rocks rising out of the fields and nestled here and there in the woods are old sea stacks that once stood proud in the surf. Except for the old sea stacks, there is no bedrock along the highway because the terrace surface is covered by sand and mud laid down along the beach and offshore while the area was underwater. There are good outcrops in the modern sea cliffs where waves are slowly cutting away the old terrace.

One of the best known of those old sea stacks that rise above the marine terrace was Tupper Rock which used to be a prominent landmark at Bandon. Unfortunately for its career as a landmark, Tupper Rock was made of blueschist which is the best of all possible rocks for rip rap and jettystone because it is heavy and exceedingly difficult to break — you need a big hammer to collect the stuff. Waves don't easily move or break rip rap made of blueschist. So Tupper Rock was quarried right off the map and most of it is now in the south jetty at Bandon which is one of the best places in Oregon to see nice, fresh pieces of blueschist. The jetty at Coos Bay is another good blueschist locality but that came from another quarry.

Here and there the sediments blanketing the marine terrace include deposits of black sand that contain as much as 10 per cent chromite along with small amounts of platinum and gold. Both the chromite and platinum come from patches of old oceanic crust in the Klamaths, from dense, black igneous rocks that usually contain these minerals. Gold never comes from such rocks but there are some lighter-colored igneous rocks in the Klamaths which have deposits of gold associated with them. Streams bring all these minerals from their various sources to the coast where the waves sort them, because they are all heavy, into deposits of black sand. Several small mines have worked the black sands in this area during times when chromite was in short supply and several have attempted to work them for platinum and gold. But reserves are small so future prospects are bleak.

Somewhere in the tangle of hills about 20 miles east of Port Orford is an enormous meteorite found by Dr. John Evans, an early Oregon explorer, back in July, 1856 and never seen since. There is no doubt that it exists because Dr. Evans sent a specimen to the Smithsonian Museum which still has it. He estimated that the entire mass must weigh more than 10 tons which, if true, would place it among the ten largest metoerites ever found. Hundreds of people have looked for the lost Port Orford meteorite and there are dozens of theories about where it might be, but most students of the problem seem to think that it probably lies somewhere in the headwaters of the Elk River, possibly on the west slope of Iron Mountain. The meteorite, if found, will make an interesting museum specimen but will not bring untold riches to its discoverer as so many searchers have imagined. We know from the specimen in Washington that it is composed mostly of an alloy of iron and nickel and has crystals of a green mineral called olivine scattered through it.

oregon 18

lincoln city — portland

The route between Lincoln City and Portland crosses both the Coast Range and the Willamette Valley, passing some very interesting rocks which are hard to see. Thick soils and all that greenery keep western Oregon's rocks remarkably well hidden.

Lincoln City is on an old beach now raised above the waves, either by uplift of the land or a drop in sea level, to make a smooth and nearly level terrace overlooking the modern beach; an excellent townsite. The route follows this terrace to Neotsu where highway 18 turns inland, crossing about 3 miles of mudstones laid down on the seafloor during Eocene time, about 50 million years ago. The road follows the Salmon River inland from Otis, passing through Eocene seafloor basalts for about 6 miles and then back into Eocene mudstones all the way through Grande Ronde to Willamina.

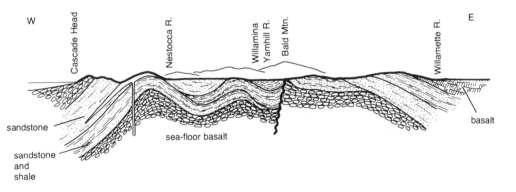

Section through Coast Range just north of Lincoln City.

From Willamina the highway follows the broad floodplain of the Yamhill River through McMinnville to Newberg on the Willamette River. Just east of Newberg the road leaves the floodplain to cut

18
LINCOLN CITY — PORTLAND
(70 miles or 113 kilometers)

22
SALEM — HEBO JCT.
(55 miles or 89 kilometers)

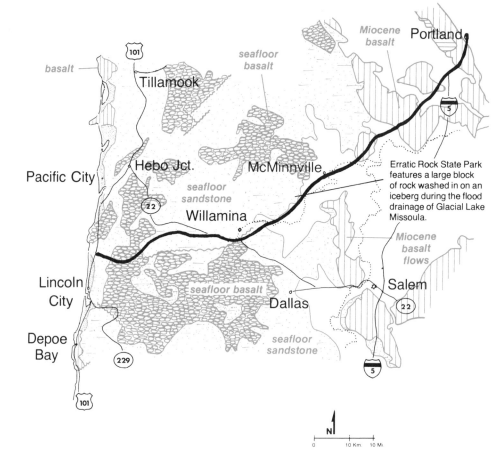

Miocene basalt

seafloor basalt

basalt

Tillamook

Hebo Jct.

Pacific City

seafloor sandstone

McMinnville

Willamina

Portland

Erratic Rock State Park features a large block of rock washed in on an iceberg during the flood drainage of Glacial Lake Missoula.

Miocene basalt flows

Lincoln City

seafloor basalt

Dallas

Salem

Depoe Bay

seafloor sandstone

N

0   10 Km.   10 Mi.

A landslide in the wave-eroded edge of an old coastal terrace threatens homes in Lincoln City.

across a tract of low hills eroded on basalt lava flows which it follows into the outskirts of Portland. These basalt flows are the westernmost end of the Columbia Plateau; they poured all the way from northeastern Oregon down the Columbia River and onto the north ends of the Coast Range and Willamette Valley. Those eruptions happened during Miocene time, about 20 million years ago, so they are much younger than the rocks in the Coast Range.

The line of hills which runs just west of Portland through Sylvan, Oswego and West Linn is a chain of small volcanoes which were active during the last several million years. There are literally dozens of small volcanoes around Portland, grouped in clusters all around the town.

38
INTERSTATE 5 — REEDSPORT
(57 miles or 92 kilometers)

42
ROSEBURG — BANDON
(84 miles or 136 kilometers)

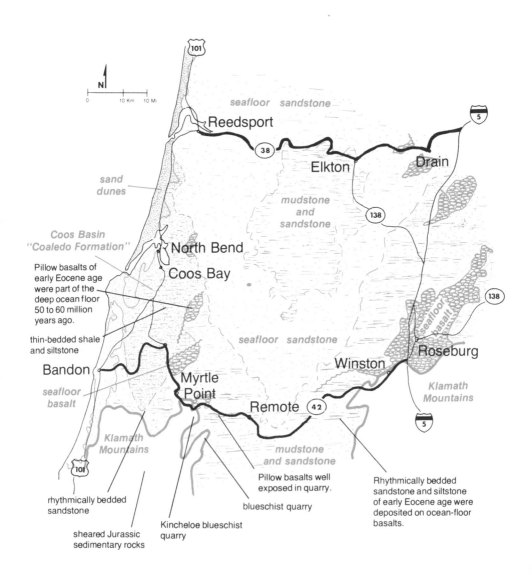

N

0   10 Km   10 Mi

seafloor   sandstone

Reedsport

38

Elkton

Drain

101

5

*sand
dunes*

138

*mudstone
and
sandstone*

**Coos Basin
"Coaledo Formation"**

North Bend

Coos Bay

Pillow basalts of
early Eocene age
were part of the
deep ocean floor
50 to 60 million
years ago.

thin-bedded shale
and siltstone

seafloor   sandstone

*seafloor
basalt*

Winston

Roseburg

138

Bandon

Myrtle
Point

Remote

42

*Klamath
Mountains*

5

*seafloor
basalt*

*Klamath
Mountains*

101

*mudstone
and sandstone*

Pillow basalts well
exposed in quarry.

rhythmically bedded
sandstone

blueschist quarry

Rhythmically bedded
sandstone and siltstone
of early Eocene age were
deposited on ocean-floor
basalts.

sheared Jurassic
sedimentary rocks

Kincheloe blueschist
quarry

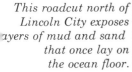

This roadcut north of Lincoln City exposes layers of mud and sand that once lay on the ocean floor.

oregon 38

reedsport — i-5

Oregon 38 follows the broad valley of the Umpqua River and then the narrower valley of Elk Creek through gentle green hills, a beautiful and sparsely inhabited part of the Coast Range.

Except for a small patch of basalt about 8 miles west of Drain, every rock along the way is mudstone or dirty sandstone deposited on the floor of the Pacific Ocean during Eocene time, perhaps about 50 million years ago. None of these rocks are well exposed, partly because they are very soft and partly because these hills, like most of the Coast Range, are covered by deep soil and dense forests.

The basalts west of Drain, and another larger patch of basalt just off the road immediately south of Drain, are part of the old bedrock seafloor on which the sediments were deposited. Basalt is a much harder rock than the sandstones and it more stoutly resists erosion. Mt. Yoncalla just south of Drain is an erosional remnant of basalt.

It always seems surprising that the lush landscape of the Coast Range supports so few people. The reason is that the soils, although deep, are very poor and deficient in almost every natural fertilizer nutrient. The rainy climate of the Coast Range washes the soluble fertilizer nutrients away leaving a sterile residue of insoluble iron oxides and clay. In many places the tropical lateritic soils that formed during Miocene time still survive. They are the most impoverished soils of all.

Fertilizer doesn't cure these soils because they contain a kind of clay that won't retain nutrients. On the atomic scale, all clay minerals are made of thin layers, resembling mica flakes, stacked like the pages of a book. In most soils the clays expand and absorb fertilizer nutrients between the layers, holding them there and then releasing them to plant roots. The clays in the red soils of the Coast Range don't do this, so fertilizer washes right through with the rain.

Trees grow well in such sterile soils because they require very little fertilizer. But food crops do need fertilizer nutrients so they languish and the little they produce is not nearly so nutritious as it would have been had it grown on better soil. Trees are the only crop that will ever flourish in the Coast Range and forestry the only kind of agriculture that will prosper.

oregon 42

bandon — i-5

Between Bandon and I-5, Oregon 42 passes all the way through the Coast Range into the Klamaths over quite a variety of unusually interesting rocks. Exposures along much of the route through the Coast Range are excellent, better than those along any other good road through these mountains.

Between Bandon and Coquille the road crosses the southern end of the Coos Basin and then between Coquille and Remote it crosses a thick sequence of sandstones and mudstones that are crumpled into tight folds, cut by numerous large faults, and mixed in with prominent slices of the basalt bedrock of the ocean floor. Everything about these rocks suggests deposition in very deep water. East of Remote the road passes through more sandstones and mudstones which are only slightly deformed, occasionally show signs of deposition in shallow water, and contain no slices of seafloor. All of these rocks are about the same age; they were deposited during Eocene time, between 50 and 40 million years ago.

It seems likely that the rocks along Oregon 42 are the remains of an old continental margin that telescoped together as the slow landward movement of the Pacific Ocean floor jammed it into the edge of the Klamaths late in Eocene time. The severely deformed rocks between Coquille and Remote appear to be sediments dumped onto the seafloor in deep water, and the less deformed ones farther east the old continental shelf. After these rocks were crushed into the Coast Range, the younger Eocene rocks in the Coos basin west of Coquille accumulated on top of them.

The area between Bandon and Coquille is at the southern end of the Coos basin, a large fold in the Coast Range that extends about 25 miles from north to south and is about 12 miles across from west to east. Most of the area within this trough-shaped fold is underlain by the Coaledo Formation, a sequence of some thousands of feet of sandstones and mudstones deposited along the shoreline of what must have been a shallow bay late in Eocene time, perhaps 45 million years ago. Both the upper and lower parts of the formation contain thick coal seams that supported rather large mines for many years. Mining began in 1854 and production reached a peak of about 100,000 tons annually during the years around the turn of the century, most of it going to San Francisco on sailing vessels. But Oregon coal mines could not compete with the California oil wells which began to flow in 1898. Coal production dropped to almost nothing by 1920 and ceased completely during the second world war. The old mines are now so overgrown that they are difficult to find.

Only a minute fraction of the available coal was mined; estimates of the remaining reserves in the Coos basin vary widely but range as high as 51 million tons. The quality of Coos basin coal is uniformly poor because it contains a lot of water and ash and relatively little heat. But it is coal, and it is useable, and fossil fuel resources of all kinds are becoming scarce. We will probably see more coal mining in the Coos basin.

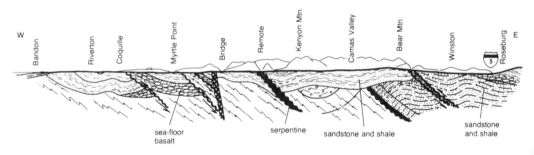

Section along the line of Oregon 42 across the Coos coal basin.

Between Coquille and the area about 2 miles south of Myrtle Point, Oregon 42 follows the eastern edge of the broad floodplain of the Coquille River. The crumpled rocks in this area were all once part of the Pacific Ocean floor, either the hard basalt lava flows that make the upper part of the bedrock seafloor or the layers of brown sandstone and mudstone that later covered them. There are large roadcuts in the dark basalts just a few miles east of Coquille.

"Pillow basalt" exposedin the wall of a roadside quarry 2 miles west of Bridge Junction. Basalt lava flows erupted under water always look like this.

Between Myrtle Point and Bridge Junction the road crosses two tracts of old seafloor basalts, both easy to spot because there are large roadcuts in very dark rock. In the easternmost of these, just 2 miles west of Bridge junction, a large quarry north of the road provides an exciting view of the seafloor. As usually happens when flows erupt underwater, the basalt extruded, like toothpaste, into long cylinders several feet in diameter. These look like stacks of black sofa pillows where they are cross-sectioned in the quarry walls and like they were still lying on the seafloor itself where they litter part of the quarry floor. The lower part of the quarry cuts through the basalt, the upper part through dirty gray sandstone deposited on top of it. The upper surface of the basalt flows is very irregular with rough crags of basalt rising into the sandstone. It is the rugged bedrock floor of the Pacific Ocean preserved under a blanket of sand just as it formed millions of years ago and hundreds of miles away in deep water at the old mid-ocean rise.

These cylinders of basalt littering the floor of a quarry beside the road 2 miles west of Bridge Junction are basalt "pillows." They are several feet in diameter and look exactly like basalt photographed in modern mid-ocean ridges by research submarines.

107

Massive beds of muddy sandstones make bold outcrops beside the road east of Remote.

There is a blueschist quarry just south of Bridge junction. Evidently a large fault must pass through this area, although none is shown on geologic maps, because blueschists form under very high pressures deep below the surface and then return, carried along in masses of serpentinite which follow faults as they squeeze back to the surface. The road does cross a little patch of serpentinite about 2 miles east of Bridge junction.

East of Bridge junction the road follows the Middle Fork of the Coquille River all the way to its headwaters. There are magnificent outcrops of dirty brown sandstone all along the way, rocks that are common throughout most of the Coast Range but rarely so well exposed. They give an excellent impression of the kind of sediment that probably forms most of the continental shelf.

Between a few miles west of Winston and I-5, Oregon 42 just barely nips across the northernmost end of the Klamath Range. But this really is a geologic technicality because there is nothing in either the landscape or the rocks along the road to give a clue.

Roseburg is in the eastern edge of the Coast Range among low hills eroded into basalt lava flows that used to be on the floor of the Pacific Ocean. It is right in the corner where three geologic provinces meet; the hills south of town are in the Klamaths, those east of town in the western Cascades, and those west of town in the Coast Range.

the cascades

The Cascades are a complex volcanic chain with a history of intermittent eruptive activity that goes back about 50 million years to Eocene time, and continues today. Geologists broadly divide the range into two main parts: the western Cascades composed of older rocks erupted from volcanoes extinct for millions of years, and the high Cascades composed of active or very recently extinct volcanoes.

The oldest Cascade volcanic rocks lie along the westernmost fringe of the range south of Eugene. They mesh along the edge of the Willamette Valley with sedimentary rocks deposited along the seashore and with other sediments deposited on land. Both contain fossils—plants and animals that lived during that same period.

At exactly the same time, similar rocks were erupting from a group of volcanoes in the Ochoco Mountains of central Oregon. Like the Eocene rocks in the western Cascades, those in the Ochocos are andesites and basalts which also contain abundant plant fossils. However, they don't contain any rocks deposited along the seashore.

This arrangement of rocks in time and space poses some very interesting questions. Evidently the seashore was right at the foot of the southern part of the western Cascades when they began their career — no doubt about that. But it does seem very strange that the earliest western Cascade volcanics should be confined to the part of the range south of Eugene and that exactly similar rocks of exactly the same age should appear again in the Ochoco Mountains.

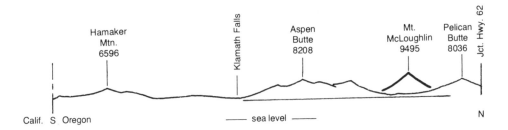

Volcanic chains similar to the Cascades normally run parallel to a coastline and the most reasonable solution to the problem is to imagine the Eocene coastline curving eastward just north of Eugene and passing east through northern Oregon somewhere north of the Ochocos. So the large ocean bay that existed earlier in much of central Oregon still existed in Eocene time but had grown much smaller.

Viewed in this way, the Ochocos are actually, in a geologic sense, part of the early Cascades. A continuous belt of Eocene volcanics probably exists in the area between Eugene and the Ochocos buried under younger rocks. Rocks similar in age and appearance to those in the Ochocos show up here and there as far east as Pendleton, so we believe that the Eocene embayment probably extended nearly that far east.

After the earliest episode of volcanic activity in the western Cascades had ended, there was a long period of volcanically uneventful quiet. The next sequence of eruptions occurred during late Oligocene and early Miocene time, between about 30 and 20 million years ago. These younger rocks extend the entire length of the Oregon Cascades so the coastline had by this time straightened itself and the great embayment that had for so long existed in central Oregon was finally gone.

This interpretation is one of those that raises as many questions as it answers. It requires that the old line of sea floor sinking pass just north of the Klamaths and curve northeastward through central Oregon to the area around Walla Walla. The problem is that such a line of sea floor sinking passes within a few miles of the southern part of the western Cas-

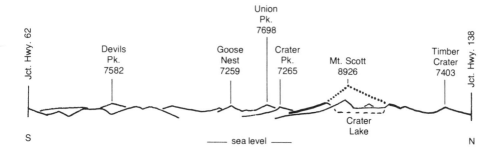

cades. They should be separated by about 100 miles. We aren't sure how to solve this problem except to suggest that it may exist simply because the map of Oregon has been stretched considerably out of shape during the past 35 million years by the persistent northward movement of the western part of the state. We can't be sure just how great this distortion may have been because so many of the rocks that might help answer that question are buried under younger volcanics so we used a thoroughly conservative estimate in preparing our own maps. If the amount is somewhat greater than what we depict in our maps, the problem would disappear. The rather sharp bend in the line of the older volcanic chain certainly suggests some considerable distortion because such chains normally form smooth curves.

This second major episode of volcanic activity in the western Cascades produced more dark lava flows as well as very large volumes of light-colored rock, the sort that usually erupts as ash. There are large deposits of volcanic ash of the same age and composition in central Oregon, called the John Day Formation, which probably blew in from the western Cascades.

The third and last major period of volcanic activity in the western Cascades began towards the end of Miocene time, perhaps 10 to 15 million years ago. These eruptions produced mostly darker rocks and relatively little light-colored ash. These eruptions, like those in the period just preceding, went the entire length of the Oregon Cascades.

It is possible to recognize old volcanic vents by the kinds of rocks they contain long after the slow processes of erosion have

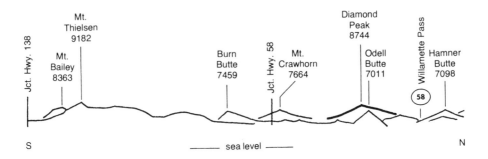

carved the original volcano right off the landscape. No one knows ₁.ow many old vents may exist in the western Cascades but at least several dozen are now located and their distribution shows a crude but clear pattern: they form a series of roughly parallel north-south lines which are oldest in the west and become progressively younger eastward. Evidently each successive phase of activity formed a new row of volcanoes one step east of the last one.

The last eruptions in the western Cascades ended about 12 million years ago, at the end of Miocene time. The processes of erosion are painfully slow but 12 million years is a long time and nothing remains of the original volcanic landscape. Hardly anything in the modern ridge and ravine landscape of the western Cascades suggests that this is a volcanic range. The bedrock is volcanic but the mountains are not volcanoes — they are simply erosional mountains carved in volcanic rock and they look about the same as erosional mountains carved in any kind of fairly uniform bedrock.

Of course erosion is not the only reason for the lack of volcanic cones in the western Cascades. Another is the fact that each new episode of activity tended to bury the older volcanoes under younger rocks. So the volcanic complexity of the range, as well as erosion, contributes to the difficulty of recognizing old volcanoes in the western Cascades. In the Ochoco Mountains where there was only one episode of activity, during Eocene time, it is still possible to recognize some of the old volcanoes as individual mountain peaks.

Sometime after volcanic activity ceased, the entire western

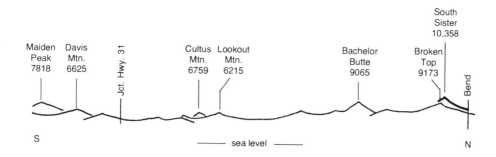

South
Sister
10,358

Maiden   Davis
Peak   Mtn.
7818   6625

Jct. Hwy. 31

Cultus Lookout
Mtn.   Mtn.
6759   6215

Bachelor
Butte
9065

Broken
Top
9173

Bend

S

———— sea level ————

N

Cascade range subsided very slightly folding the rocks into a broad structural trough that runs the length of the range. This very gentle fold in the rocks is not expressed in the landscape and certainly is not the sort of thing that anyone would notice while driving through. It is a subtle structure that remained hidden until geologists had separated the rocks belonging to the three major volcanic sequences and plotted their distribution on geologic maps. The map pattern quite clearly shows the fold.

No one can be quite sure why the western Cascades are slightly sunken into a broad structural trough but it seems very likely that the subsidence along the length of the range is due largely to cooling of the rocks beneath. It is interesting and probably significant that the modern high Cascades appear to be perched along the crest of an equally broad and subtle arch, perhaps because the crust beneath them is hot and still charged with molten magma.

Volcanic rocks come in several varieties which usually betray themselves by their colors. Basalt is always black unless some kind of weathering or alteration has stained it red or green and it always erupts rather quietly, tending to pour fluidly out on the surface to make big lava flows. Large basalt volcanoes are shaped about like giant vanilla wafers and they don't cut much of a profile against the skyline even when they are enormous. In proper scientific terms these are called shield volcanoes because of a fancied resemblance in shape to that of a Roman shield — a comparison that doesn't seem too vivid now that it has been nearly 2000 years since Roman shields were familiar objects.

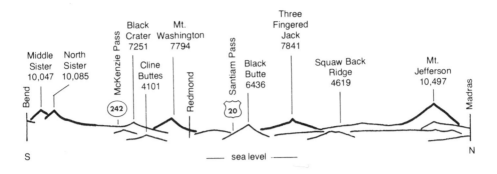

Rhyolite is at the other extreme. It is always very pale, usually some light shade of gray, pink, or yellow, and the molten magma is often so charged with steam that it blasts violently out of the vent as towering clouds of hot ash. Sometimes rhyolite actually explodes. Rhyolite magmas are too stiff and pasty to pour easily so if they don't happen to contain much steam they are likely to extrude through the surface as a mountain of magma, properly known as a plug dome, that rises as though it were a giant puffball. If they are very dry, they make obsidian lava flows.

Andesites, the most typical Cascade rocks, are intermediate between basalt and rhyolite in appearance, composition, and behavior. They are usually some shade of gray, brown, or green. Sometimes they erupt quietly, like basalt, to make lava flows and at other times they may fill the sky with clouds of ash. Andesite eruptions build the tall, symmetrical cones that picturesquely punctuate the skyline and help maintain prosperity in the photographic supply business.

The latest phase of volcanic activity which built the modern high Cascades appears to have begun several million years ago and still continues. The first magmas were basalts which built a continuous row of massive volcanoes along the crest of the broad structural arch that runs the length of the high Cascades. These early basalt volcanoes are quite substantial but, like all large basalt volcanoes, they have low profiles. So they raised the skyline some thousands of feet without giving it much character. Some of the big basalt volcanoes are still active; Belknap between McKenzie and Santiam Passes is a good example. After the long ridge of coalescing basalt vol-

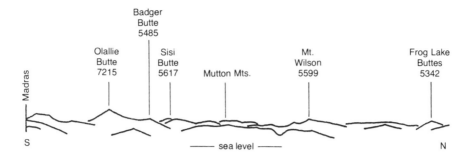

canoes had already become quite substantial, andesite magmas began to build the imposing chain of towering Cascade Peaks upon a foundation of older basalts.

Several of the high Cascade volcanoes, from the Three Sisters south, have now progressed beyond erupting ordinary andesite and begun to produce really light-colored rocks, even rhyolites in some cases. None of the large cones in Oregon north of the Three Sisters have yet produced anything lighter than andesite and neither are there very many light-colored rocks north of that area in the older volcanics of the western Cascades. The step from andesite to rhyolite is important because it usually causes a change in the volcano's style of eruption. And the distribution of andesite and rhyolite may also tell us something about the kinds of rocks that lie beneath the cascade volcanics.

All volcanic eruptions, even the relatively quiet basaltic ones, are impressive fireworks displays which can be dangerous enough to anyone who gets too close. Andesitic eruptions are very likely to produce enormous clouds of volcanic ash which may blanket thousands of square miles of the surrounding countryside. And they are also likely to produce devastating mudflows. Freshly fallen volcanic ash makes wonderful mud if it mixes with a bit of rain or melting snow and the mud, being much denser than water, can carry large boulders. Big mudflows pouring down the flanks of an andesite cone during and after an eruption are extremely dangerous, probably the greatest threat these volcanoes pose. But a large rhyolite eruption is likely to develop into a general holocaust, the most devastating kind of natural disaster. Rhyolite looks innocent

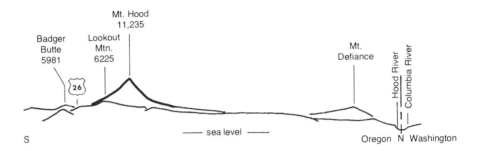

enough once it has cooled but the molten magma has real destructive potential.

Rhyolite melts to produce a stiff magma with a consistency resembling that of kindergarten modelling clay. As long as it stays dry, rhyolite magma is relatively harmless and will erupt quietly to make thick obsidian lava flows or rhyolite plug domes. But if it gets a chance to absorb some water, and rhyolite is capable of absorbing much more water than either basalt or andesite, the volcano becomes a giant steam boiler at a temperature of about 850° centegrade, the melting point of rhyolite. The only thing holding it down is the weight of the overlying rock and once such a volcano begins to erupt the lid is off and the entire magma chamber may vomit through the vent in a blast of steam and shredded rhyolite magma that will blow clouds of ash high into the sky and cover the nearby countryside with a blanket of steaming ash and pumice. Then the volcano subsides into the emptied magma chamber beneath leaving a large crater at the surface. Mount Mazama did this a little more than 6000 years ago leaving a hole now filled by Crater Lake.

Geologists cast a suspicious eye at the other big andesite cones that have recently produced small amounts of rhyolite wondering if they may be cooking up a large mass of steamy rhyolite at depth and getting ready to blow it off. There is no way to know what some of those volcanoes may be brewing and no way to predict what they may do but several of the big peaks in the southern Cascades probably have the potential.

Several Cascade volcanoes have erupted during historic

116

time, the most recent performance being that of California's Mount Lassen in 1916. Early explorers and settlers in the Pacific northwest left a number of accounts of Cascade eruptions, some of them convincing and others not. For example, there is one report of an eruption on Mount Olympus in Washington which isn't even a volcano. So it is difficult to be sure which of the old newspaper stories actually describe real eruptions and which are merely records of forest fires, unusual clouds, or perhaps outright inebriation. Mount Hood is the only Oregon volcano mentioned in old newspapers as having been active and the reports are not convincing — it certainly didn't do much, if anything. Reports of activity in Mount Rainier are likewise unimpressive and we can be sure that nothing much happened there either. But there is no doubt whatever that significant eruptions did occur near Mount Lassen about 1851 and on several occasions on Mount Baker and Mount St. Helens.

Mount St. Helens, in Washington just across the Columbia River from Mount Hood, is probably the most active volcano in the Cascades. It erupted several times in the last century, the most recent event being a fairly vigorous outburst in 1857. Most of the volcano is less than 3000 years old and its entire upper part is only a few hundred years old. Its record of activity during recent centuries suggests that we may not have to wait too many years to see more activity in Mount St. Helens.

However, the historic record is much too short to provide reliable information on which Cascade volcanoes are still active and which certainly extinct. It is perfectly possible for an active volcano to remain quiet for hundreds or even thousands of years between eruptions and the recorded history of the Cascades goes back less than two centuries.

The appearance of a volcano is one of the best guides to whether or not it may still be active. Any large andesite volcano which is smoothly conical and shows little evidence of dissection by streams or glaciers should be regarded as probably active whether or not there is any historic record of eruptions. Mount McLoughlin near Klamath Falls is an excellent example of such a peak. Any deeply eroded volcano is probably extinct. Three Fingered Jack and Diamond Peak are both excellent examples; neither has repaired the wounds inflicted by ice age glaciers, so obviously neither has erupted since the last ice age ended, about 10,000 years ago.

We wrote the preceding paragraphs during the fall of 1976 for the first printing of this book. When Mount St. Helens actually started to erupt on March 27, 1980, it confirmed a geologic prediction based on normal methods of geologic observation. We hasten to add that we had no special knowledge; that our assessment was based on nothing more than the volcano's record of vigorous activity until 1857 and its general appearance of having grown during the very recent past.

The activity actually began on March 20 with flurries of small earthquakes, most of which were felt only very locally in the general vicinity of Spirit Lake on the northern flank of the mountain. They aroused some considerable suspicion because earthquakes generally shake large areas except in the vicinity of volcanoes on the verge of erupting. However, there appeared to be no cause for great anxiety because such earthquake swarms often turn out to be false alarms. In any case, the beautifully symmetrical shape of Mount St. Helens suggested that it probably did not have a history of great violence.

On March 27, the volcano began to blow off steam and clouds of dark ash from a small crater newly opened in its snow covered summit. However, microscopic study showed that the ash appeared to be old rock rather than new magma freshly arrived at the surface, and the steam was not very hot by volcanic standards. Evidently magma was rising deep beneath the volcano and the rain and snowmelt water that had soaked into it boiled back to the surface as steam. Many volcanos do that sort of thing from time to time without actually going into major eruption.

The clouds of ash and steam had diminished in both size and frequency by mid-April, as had the earthquakes, and it appeared then that the show might be ending except that the north side of the mountain had begun to bulge. The swelling continued at an increasing rate during the next several weeks and it was clear to everyone that a large mass of magma was pushing into the mountain. That suggested several possibilities which depended upon the steam content of the magma.

If the magma were reasonably dry, it would probably erupt as an andesite lava flow or rhyolite plug dome, neither of which would cause undue commotion. However, a dry eruption seemed unlikely because Mount St. Helens consists largely of ash and appears to have produced very few lava flows. If the magma were wet, it might crystallize quietly within the volcano without breaking the surface or it might explode into a cloud of ash. In any case, it seemed likely that the magma was relatively dry because the bulging north flank of the volcano was not heating up as fast as we might expect of a mountain full of steaming rock. The volcano gave no warning that anyone could interpret in advance.

On the morning of May 18, it exploded with a blast fully comparable to that of a 10 megaton hydrogen bomb, blowing a filthy cloud of ash more than 12 miles into the sky. The energy behind the explosion was steam, steam at a temperature of about 1200° Fahrenheit, steam hot enough to glow red in the dark. The roar of the explosion was heard as far away as the mountains of western Montana where it sounded like the rumble of distant thunder. The ash came later.

The black cloud of ash spread directly eastward as it blew downwind like a plume of smoke from a chimney. The cloud suddenly appeared in the western sky as a dark pall that quickly snuffed out the sun and then began dropping powdered rock from the sky as though it were snow. By evening it had spread across northern Idaho and into western Montana, covering everything beneath it with a blanket of dust that had the color and texture of portland cement. On Monday morning the people who lived in the track of the ash cloud awoke to a landscape drained of all color and communities in which all activity had ground to a stop. There was nothing to do but clear the ash in any way possible and hope for drenching rains.

For some reason that we cannot explain, volcanic ash kills bugs. The ash fall cleared the air of flying insects, especially of bees, and also of the birds that feed on them. Several weeks passed before butterflies and swallows came back to the area where the ash fell and the landscape seemed curiously empty without them. However, there is no reason to believe that the ash is acid, as many people feared, or in any way poisonous. It consists simply of minute particles of rock which differ from ordinary dust particles mainly in containing numerous sharp edges which make them very harshly abrasive. Maybe volcanic ash kills bugs simply by cutting them.

Fortunately, a series of good pictures were taken of Mount St. Helens just as it began to explode on the morning of May 18. They show an enormous landslide moving down the mountain as the eruption cloud boils upward out of its northern flank. We suspect that the slow bulging of the northern side of the mountain finally started the landslide which in turn triggered the eruption by relieving pressure on the magma within. The effect was like uncorking a bottle of warm champagne; all the gas in the magma, in this case steam, suddenly expanded, blowing it out of the mountain as a cloud of ash. High powered microscopes show that every minute particle of ash is full of tiny bubbles which must have formed when the magma suddenly expanded at the moment of the explosion. The particles of ash are bits of glass foam; it is the sharp edges of all the little broken bubbles that make it so abrasive.

The big blast of May 18 opened an enormous crater in the northern side of Mount St. Helens which formed partly as a result of the landslide that accompanied the explosion, partly as a result of the explosion itself, and partly because the top of the mountain simply collapsed into the void left by the magma blown out. The entire top of the volcano simply disappeared. In the moment of explosion, Mount St. Helens lowered its elevation by some 1200 feet and dropped from being the fifth highest peak in Washington to a rank somewhere near the thirtieth.

Mount St. Helens produced two later explosions neither of which were nearly as violent as the incredible blast of May 18. However, they did provide many more people with the experience of a volcanic ash fall, first in the coastal areas of Washington and next in the part of Oregon that extends from Portland southwest to the vicinity of Tillamook. All three of the large blasts came within a few days of the full moon or new moon, the times when the sun and moon exert their greatest tidal pull on the earth. It is true that volcanos do show a slightly greater tendency to erupt at those times but it would be a mistake to

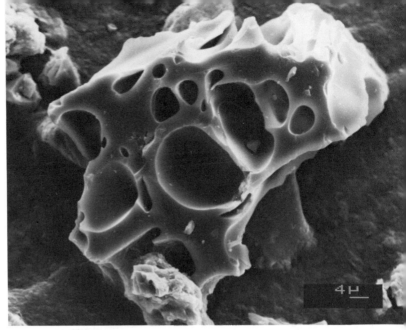

A fragment of ash as it appears magnified 1,500 times in a scanning electron microscope. The holes in this speck of glass are gas bubbles created by steam expanding within the lava at the moment of the eruption. Notice all the sharp edges.

suppose that the sun and moon have more than a very minor influence on their behavior. The tides may trigger an eruption that was about to happen in any case but they do not cause volcanos to erupt.

By the middle of June, a large lava dome had begun to grow in the new crater, expanding from one day to the next as though it were a gigantic puffball mushroom. Evidently only the top of the rising mass of magma was charged with steam. That part blew off as ash in the series of large explosions, leaving the much drier magma below to rise quietly in the crater. It is very viscous material with a composition approaching that of rhyolite so it will form a dome-shaped exrusion rather than a lava flow.

Observation of other volcanos generally similar to Mount St. Helens shows that emplacement of a large lava dome in their craters seems to plug their plumbing permanently. The main volcano becomes extinct and future activity produces a rash of cinder cones and plug domes about its flanks. Therefore, having begun this section with the confirmation of a prediction, we will end it by going out on a limb with a new prediction. If other volcanoes are any guide, Mount St. Helens will never again produce a major eruption once the current period of activity has ended. It may blow off clouds of steam and ash from time to time as water soaking into the mountain encounters the hot rock still within but it will never again erupt with the violence of the early summer of 1980. However, the Cascade chain contains many other peaks which appear to be fully as capable of producing a major eruption as Mount St. Helens. We think the most likely candidates are Mount Garibaldi in British Columbia, Mount Baker and Mount Adams in Washington, the McKenzie and Santiam Pass area, Newberry Volcano, and Mount McLoughlin in Oregon and Mount Shasta in California. There are others.

The Cascades exist because the Pacific Ocean floor has been sliding beneath the North American plate just off the Oregon coast for the past 50 million or more years. Of that there is no doubt. During Eocene time the zone of sinking must still have looped eastward following the margins of the great embayment of the ocean that has since become central Oregon. But then the zone of sinking ended to the east and began along the Coast Range. Since then the Pacific Ocean floor has been sinking along a line close to the present coast.

The cold sinking slab of seafloor heats up as it descends into the earth's interior, finally getting hot enough to begin melting the basalt that forms its bedrock crust. Molten basalt is lighter than other rocks in the earth's interior and is also very fluid and mobile so the magma rises to the surface to erupt through volcanoes. Thus the Cascades.

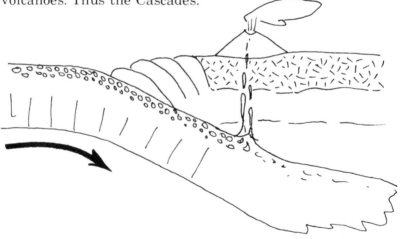

Diagrammatic section showing molten basalt rising from a sinking slab of seafloor as it gets hot enough to melt basalt.

The Cascade volcanic chain is parallel to the zone of sinking but nearly 100 miles inland from it because the sinking slab of seafloor slopes that far into the earth's interior before its basaltic crust begins to melt. It would be extremely interesting to understand why the line of active volcanoes has stepped eastward with each successive major period of activity starting with the western Cascades and culminating in the modern high Cascades. The zone of sinking has certainly not moved eastward in that time — westward if at all. Perhaps the descending slab sank more steeply in times past so it didn't get so far inland before its basaltic crust melted. No one seems to be sure. It is true, though, that this kind of shifting of volcanic chains over a long period of time seems to be typical and not peculiar to the Cascades.

Why did the Cascades go through several major periods of activity separated by intervals lasting millions of years during which nothing volcanic happened? It is difficult to imagine how a sinking slab of seafloor could cause such periodic activity if it were descending at a constant rate and even more difficult to imagine how such a slab could start and stop. Many geologists speculate that perhaps the end of a sinking slab of seafloor may break off and slide down ahead of the rest as the soft rocks of the earth's interior flow in to fill the gap behind it. If such a thing did happen, there would have been periods when no old seafloor was passing beneath the Cascades and presumably those would have been times of volcanic quiet.

A good bit of the magma erupted from Cascade volcanoes has been basalt, almost certainly part of the old basalt of the seafloor remelted in the earth's interior and recycled back to the surface. But more of the Cascade volcanics are andesites and lighter-colored rocks, even rhyolites. It is impossible to imagine how these could come directly from the sinking slab of seafloor.

Mount Hood has a plug dome in its top and some geologists would interpret that as evidence that the volcano is dead.

Andesite and the various other lighter-colored volcanic rocks differ from basalt in several ways the most important of which is their much higher proportion of silica — the more silica the lighter the rock. Molten basalt has a much higher melting point than any of the lighter-colored magmas so it is perfectly possible for it to melt and mix with some silica-rich rock to produce andesite and perhaps even rhyolite magma. Most of the common sedimentary sandstones and mudstones contain a high proportion of silica so basalt can convert itself into a lighter-colored magma by melting and assimilating them.

It seems almost certain that this must happen somewhere fairly near the surface, not deep within the interior of the earth on the descending slab of seafloor. Although the seafloor usually does carry a thick blanket of sandstone and mudstone,

124

these can hardly sink into the interior with the other rocks because they are much too light. Sinking sandstone and mudstone into the earth's interior would present a problem similar to sinking a marshmallow into a cup of cocoa. The lighter sedimentary rocks stay behind in coastal mountain ranges.

So we believe that andesite and rhyolite magmas must form when molten basalt, rising from great depths, melts and mixes with sedimentary rocks nearer the surface. Why then do we find andesite along the entire length of the Oregon Cascades and rhyolite only in their southern portion?

The answer to this question may have something to do with the curious fact that light-colored magmas can rise long distances only as long as they stay dry. Once they begin to absorb water, as steam, they soon reach a condition in which any drop in pressure will raise their melting point. Rising towards the surface must reduce the pressure on the magma. Once the pressure is reduced enough to raise the melting point above the actual temperature of the magma, it will freeze solid right where it is and then cool off later after it has already solidified. So a light-colored magma that contains a lot of steam can't rise very far before it freezes and is therefore most likely to solidify somewhere deep below the surface and most unlikely ever to erupt through a volcano. The only light-colored magmas that manage to erupt at the surface are those that start dry and manage to stay that way until they are very shallow. Some, those that become obsidian lava flows, are still perfectly dry when they erupt. The masses of rhyolite magma that absorb water to convert themselves into giant steam bombs must do this at very shallow depths.

We know that the northern part of the Oregon Cascades rests on the slab of oceanic crust that stopped moving when the line of seafloor sinking shifted to its present position about 35 million years ago. We see these rocks well exposed in the Coast Range. The cover of silica-rich sandstones and mudstones is relatively thin and still full of water. They contain very little rock that might turn into rhyolite if it were to melt. And we don't see many rocks lighter than dark andesites in this northern part of the Oregon Cascades.

The thick mass of crumpled sedimentary rocks in the Klamath Mountains contains plenty of silica and very little water. If they should melt and mix with molten basalt, the result would be a relatively dry light-colored magma that could rise to the surface and erupt before it solidified. So perhaps the distribution of dark and light-colored volcanic rocks in the Cascades is telling us what lies beneath them. Perhaps the northern limit of rhyolite rocks corresponds to the buried northern limit of the kinds of rocks that make the Klamaths.

The present condition of the Cascades is a bit of a mystery. Except for the eruption of Mount Lassen in 1916, nothing much has happened anywhere in the range for more than a century. Yet there is excellent evidence that the past several thousand years have been a period of intense activity here and there along the entire chain of volcanoes — from British Columbia to northern California. The question is whether the present quiet is merely a brief pause or an actual stop. Is there fire in the ashes?

Another Cascade mystery, which may or may not be parallel to the present lack of eruptions, is the absence of earthquakes. Unfortunately, there is no way to know whether earthquakes have been more frequent in the past several thousand years because they don't leave records of themselves the way eruptions do. Volcanic chains like the Cascades are typically shaken by frequent tremors caused by motion of the sinking slab of seafloor beneath. There are occasional earthquakes in the Cascades but not nearly as many as there should be.

What next for the Cascades? Another long period of complete inactivity as happened millions of years ago or merely a brief pause followed by renewed frequent activity as has happened several times in the last 50,000 years? We don't know but it seems more likely that this is only a brief pause. It would be fun to find some way to stir the ashes.

u.s. 20

albany — bend

Between Albany and Bend, U.S. 20 crosses the eastern edge of the Willamette Valley, the old western Cascades, and the modern high Cascades. All of the different ages and kinds of rock along the way are volcanic.

There isn't much bedrock to see in the Willamette Valley between Albany and Lebanon; the few exposures that do exist are basalt lava flows that belong to the Columbia Plateau. The highway crosses more plateau basalt flows as far east as the area around Sweet Home which is actually within the western Cascades.

The road meets the Santiam River a few miles east of Albany and follows it to its headwaters in the high Cascades. The floodplain pinches to an end about 6 miles east of Sweet Home and from there east the road follows a fairly narrow canyon which is the only kind of place in the heavily forested western slope where good outcrops are common.

U.S. 20 crosses the western Cascades between the area around Lebanon and that about 10 miles east of Santiam Junction. All the rocks along this section of road erupted between about 30 and 15 million years ago. Those along the highway are mostly andesites in various forms such as lava flows, ash beds and mudflow deposits.

Between its junction with Oregon 126 and the area just east of Santiam Junction, U.S. 20 crosses the north end of the Sand Mountain lava field. At least 22 small volcanoes are lined up along a neat north-south trend that extends from about one mile north of the highway to 7 miles south of it. Scattered trees growing on the flows suggest that they must have erupted several thousand years ago and

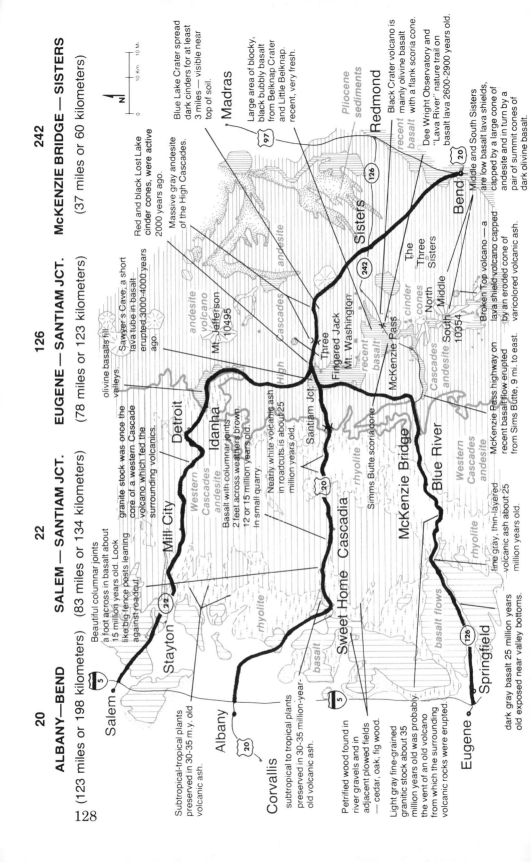

ALBANY—BEND
(123 miles or 198 kilometers)

SALEM — SANTIAM JCT.
(83 miles or 134 kilometers)

EUGENE — SANTIAM JCT.
(78 miles or 123 kilometers)

McKENZIE BRIDGE — SISTERS
(37 miles or 60 kilometers)

20   22   126   242

Blue Lake Crater spread dark cinders for at least 3 miles — visible near top of soil.

Large area of blocky, black bubbly basalt from Belknap Crater and Little Belknap. recent, very fresh.

Black Crater volcano is mainly olivine basalt with a flank scoria cone.

Dee Wright Observatory and "Lava River" nature trail on basalt lava 2600-2900 years old.

Red and black Lost Lake cinder cones, were active 2000 years ago.

Massive gray andesite of the High Cascades.

Middle and South Sisters are low basalt lava shields, capped by a large cone of andesite and in turn by a pair of summit cones of dark olivine basalt.

Broken Top volcano — a lava shield volcano capped by an eroded cone of varicolored volcanic ash.

Sawyer's Cave, a short lava tube in basalt erupted 3000-4000 years ago.

olivine basalts fill valleys.

granite stock was once the core of a western Cascade volcano which fed the surrounding volcanics.

Beautiful columnar joints a foot across in basalt about 15 million years old. Look like big fence posts leaning against roadcut.

Basalt with columnar joints 2 feet across weathers brown. 12 or 15 million years old. In small quarry.

Nearly white volcanic ash in roadcuts is about 25 million years old.

McKenzie Pass highway on recent basalt flow erupted from Sims Butte, 9 mi. to east.

fine gray, thin-layered volcanic ash about 25 million years old.

Subtropical-tropical plants preserved in 30-35 m.y. old volcanic ash.

subtropical to tropical plants preserved in 30-35 million-year old volcanic ash.

Petrified wood found in river gravels and in adjacent plowed fields — cedar, oak, fig wood.

Light gray fine-grained granitic stock about 35 million years old was probably the vent of an old volcano from which the surrounding volcanic rocks were erupted.

dark gray basalt 25 million years old exposed near valley bottoms.

Salem   Stayton   Mill City   Detroit   Idanha

Corvallis   Albany   Sweet Home   Cascadia   McKenzie Bridge   Blue River

Eugene   Springfield

Madras   Redmond   Sisters   Bend

Santiam Jct.

Mt. Jefferson 10495   Three Fingered Jack   Mt. Washington

McKenzie Pass   North Sister   Middle   South 10354

The Three Sisters

Western Cascades andesite   rhyolite   basalt

High Cascades recent basalt   andesite

Cascades andesite

Simms Butte scoria cone

Pliocene sediments

cinder cones

128

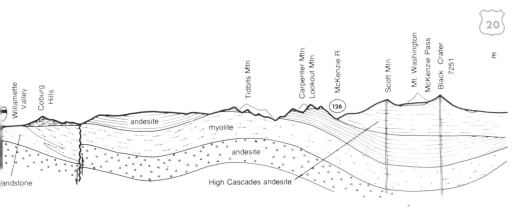

Willamette Valley · Coburg Hills · Tidbits Mtn. · Carpenter Mtn. · Lookout Mtn. · McKenzie R. · Scott Mtn. · Mt. Washington · McKenzie Pass · Black Crater 7251 · E

andesite

rhyolite

(126)

andesite

High Cascades andesite

andstone

Section north of Eugene to near Sisters. The older volcanic rocks of the western Cascades are folded into a broad trough and the modern high Cascades are piled on top.

radiocarbon dates from trees killed by one of them confirm the impression by giving a definite age of about 3000 years.

One of the earliest flows in the Sand Mountain lava field moved north from one of the volcanoes about 2 miles south of the road and ponded to form a lava lake right at Santiam Junction. After the surface of the lake had solidified, the still molten lava within drained out from beneath it, letting the solid crust sink. Ridges of basalt just south of the junction are remnants of the edge of that lava lake left standing in their original position when the center sank.

The high Cascades were thoroughly glaciated during the last ice age and the area between Santiam Junction on the west and Blue Lake on the east side of Santiam Pass was largely covered by ice. Occasional glacial features near the road are covered by a blanket of dark volcanic ash that drifted eastward on the wind during the eruptions in the Sand Mountain area.

Blue Lake crater is just south of the highway a little more than 3 miles east of Santiam Pass. It is a deep crater blasted out of solid rock by violent steam explosions. Evidently some basalt magma rising toward the surface got into some wet ground where it generated extremely hot steam and made a lot of noise and commotion without actually erupting much molten magma. Radiocarbon dates on wood from trees buried by debris settling after the explosion show that it happened about 3500 years ago. The area around the crater is littered by large quantities of shattered fragments of older volcanic rock blasted out by the explosion and small quantities of very bubbly basalt erupted when the crater formed. The gas bubbles in the fresh basalt show that it contained a lot of steam when it erupted.

About halfway between Blue Lake crater and Sisters, the road passes the south edge of Black Butte, a large cinder cone that was probably active before the last ice age. It is old enough to be well covered by trees and no longer looks fresh. The swampy area that the road crosses near the base of Black Butte exists because the volcano sits squarely on top of old streams that used to flow north through this area. Now the water backs up into the swamp, seeps through the porous volcanic rocks beneath the volcano, and reappears at Metolius Springs just north of Black Butte.

Rocks along the way between Sisters and Bend are all basalt flows sufficiently covered by soil and vegetation to be nearly invisible. They appear to be the eastern apron of the Cascades.

The route between Sisters and Bend provides wonderful views of the Cascades outlined against the western horizon. It is interesting to try to distinguish the active from the extinct volcanoes by their appearance. If the volcano is symmetrical and cone shaped, it has certainly been active since the last ice age and is very likely alive and capable of erupting again. If it is craggy, carved and bitten by glaciers, it has not been active since the last ice age and is more likely extinct.

Ice age glaciers reduced Three Fingered Jack to a few rocky crags. Such an eroded volcano is almost surely extinct.

oregon 22

salem — santiam junction (u.s. 20)

This stretch of road crosses the eastern side of the Willamette Valley and follows the North Fork of the Santiam River through the western Cascades and into the high Cascades. The drive is beautiful but trees and soil hide most of the rocks.

Between Salem and a few miles east of Stayton, the road crosses the eastern side of the Willamette Valley. This country is underlain by flood basalts that seem to be part of the Columbia Plateau but, fortunately for the economy of Oregon, soil cover is thick and basalt outcrops are few.

About four miles east of Stayton the flood basalts exposed in the valley walls above the floodplain give way to volcanics of the western Cascades. These westernmost Cascade volcanics are somewhat older and lighter in color than both the flood basalts and most of the younger rocks in the western Cascades. As the road ascends eastward to higher elevations; it gets into progressively younger rocks because the Cascades are a pile of volcanics erupted on top of each other.

A few miles east of Mill City the river floodplain narrows and finally disappears. From there upstream to Santiam Junction, the river flows along the floor of a narrow canyon and the rocks are a little more visible.

Rocks between Mill City and Idanha are almost all a complex mixture of andesite lava flows mixed in with ash beds, mudflow deposits, and the various other things that volcanoes manage in one way or another to make of their magma. Around Detroit dam, the road crosses a little granitic intrusion about a mile in diameter that nicely punctuates the surrounding volcanic monotony. This is a mass of andesitic magma that didn't make it to the surface but cooled at depth to become granite. That is likely to happen if the magma absorbs water while it is still at some depth below the surface.

All of the volcanic rocks within a broad area extending about 5 miles both east and west of Detroit dam and much farther north and south are altered by the action of hot water and steam. Similar zones of altered rock surround most of the masses of granite that intruded the volcanic rocks of the western Cascades. Hot water alteration normally bleaches rocks, in this area to various pale shades of greenish gray, and often changes them so completely that it is difficult to figure out what the original rock might have been.

We sometimes find that the same hot waters that alter and bleach rocks also load them with valuable ore minerals. Very few worthwhile ore bodies have turned up in the western Cascades despite the fact that there are several large areas of altered rock which look like promising places to prospect. The U.S. Geological Survey has carefully mapped these altered zones in an effort to guide people looking for mineral deposits. The big problem in this area, as everywhere in the western Cascades, is the dense forest and thick soil cover which hide the rocks and make prospecting difficult.

Between the Idanha area and Santiam Junction, Oregon 22 crosses more of the older, lighter-colored Cascade volcanics mostly covered by very recent lava flows from the modern high Cascades.

oregon 126

eugene — santiam junction (u.s. 20)

Springfield is at the eastern edge of the Willamette Valley, right where the younger sedimentary fill in the valley floor laps onto the older volcanic rocks of the western Cascades. All the rocks exposed along the road between Springfield and Santiam Junction are volcanic; those between Springfield and McKenzie Bridge are older volcanics belonging to the western Cascades and those farther east between McKenzie Bridge and Santiam Junction are much younger rocks erupted from the modern high Cascades.

None of the rocks are easy to see because the dense forests of the western slope cover them all with grand indifference to their age and description. Besides, the road follows the McKenzie River which has a broad floodplain along about half of the route between Springfield and McKenzie Bridge.

Western Cascade volcanics between Springfield and Leaburg are mostly very dark andesites and basalts while those between Leaburg and McKenzie Bridge tend to be lighter andesites mixed in with some very light-colored rocks, rhyolites. Both these volcanic series are very nearly the same age and both contain the usual messy mixture of lava flows, ash beds, mudflows, and other rocks typical of large volcanic mountains. However, the mountains are not the original volcanoes; those all fell victim to erosion long ago. The modern landscape is almost entirely erosional.

For a distance of about 6 miles midway between Vida and Blue River, the floodplain narrows and the McKenzie River winds through a canyon. In the middle of this canyon, along a road distance of about 4 miles, the highway cuts across the length of an oval body of coarse-grained granitic rock. The difficulty the river has had in eroding this solid rock undoubtedly explains why the narrow canyon exists.

Round or oval bodies of granitic rock a few miles in diameter are scattered along the length of the western Cascades like a crudely strung chain of beads. They are masses of andesitic magma that rose into the volcanic pile and crystallized there without actually managing to make it to the surface. Each may well be the magma chamber of an ancient volcano, now frozen into the core of the old volcanic chain.

Another and much smaller granitic mass which the road does not cross is in the hills directly north of Leaburg dam. It differs in composition from the one along the road farther east and appears to be somewhat older.

All of the large granitic plugs in the western Cascades, including the one between Vida and Blue River, are surrounded by zones several miles wide in which the surrounding volcanic rocks have been attacked and extensively altered by hot water and steam. Evidently heat from the plug of granite kept a good circulation of water moving through the nearby volcanic rocks. This is probably what is going on inside many modern volcanoes that blow a plume of steam from their summits and have steam vents and hot springs on their flanks. They just sit and stew in their own juices.

Steam and hot water are powerful solvents and no rock can withstand their prolonged attack. In the western Cascades they bleached most of the color out of the volcanic rocks in the altered zones, leaving them looking as though they might have been done by a timid artist equipped with a pale set of pastels. The circulating water and steam also bring new minerals into the altered rocks, frequently including valuable ore minerals. Nearly every mine in the western Cascades, of which there are very few, is in one of the zones of altered rock surrounding a granitic intrusion.

The floodplain ends just east of McKenzie Bridge and from there north to Santiam Junction the road follows the river upstream through a fairly narrow canyon. All of the rocks along this stretch of road are volcanics erupted from the nearby peaks of the high Cascades within the last few million years. Except for a few roadcuts in dark volcanic ash, all are lava flows of very dark andesite or basalt. Most of the country is heavily forested but some of the flows are much too young to have acquired a soil cover and the road crosses several ragged and clinkery flow surfaces.

Clear Lake, just east of the highway about 4 miles south of Santiam Junction, exists because a lava flow dammed the McKenzie River.

The lake is about 100 feet deep and still has a number of drowned trees rooted in its bottom. Radiocarbon dates on their wood show that it is about 3000 years old so that must be the age of the lava flow that dammed the river. The source of the flow is a small cinder cone volcano 3 miles east of the lake.

The numerous other natural lakes and marshes in this area all formed where lava flows disrupted the natural drainage within the past few thousand years. Surface streams are remarkably scarce, partly because lava flows have dammed them and partly because much of the drainage goes underground following the porous zones along the tops and bottoms of buried flows.

These crisp basalt columns are exposed in a roadcut beside U.S. 20 about 6 miles east of Sweet Home.

oregon 242

mckenzie bridge — sisters

The highway over McKenzie Pass goes through the most spectacular area of very recent volcanic activity accessible by road anywhere in the Oregon high Cascades. Magnificent lava flows in the top of the pass make the drive a real treat.

All the outcrops along the road are volcanics erupted from the modern peaks of the high Cascades during the very recent geologic past. Most of them are lava flows, either basalt or very dark andesite that hardly differs from basalt.

Several roadcuts around the junction of Oregon highways 242 and 126, 5 miles east of McKenzie Bridge, are in glacial moraines which mark the end of an enormous tongue of ice that flowed down the valley of Lost Creek from Middle Sister toward the end of the last ice age, perhaps 10,000 years ago. Glacial moraines are composed of an unsorted mess of all kinds and sizes of rocks jumbled indiscriminately together in a matrix of mud. Rockhounds especially appreciate these moraines because they contain beautiful chunks of obsidian brought down from Middle Sister by the ice. Look for pieces of shiny black glass along the floor of Lost Creek.

Oregon 242 winds along the valley of Lost Creek for about 12 miles east of McKenzie Bridge to Sims Butte where the route leaves the valley and turns northeast toward the pass. Sims Butte is a small volcano that poured a series of thin basalt lava flows down the valley of Lost Creek sometime since the last ice age. Several miles of road is built on these flows and they are exposed in roadcuts along some of the switchbacks. Lost Creek is so named because it and several of its tributaries disappear beneath this basalt pavement and continue by trickling through the old stream gravels buried beneath. They emerge from under the end of the basalt in several big springs which form the source of the Lost Creek that flows on the surface near McKenzie Bridge.

Linton Lake, at the head of Lost Creek canyon about a mile east of Alder Springs campground, is another volcanic creation. A lava flow from Collier cone, a volcano about 5 miles east of Sims Butte, dammed Linton Creek which tumbles into the lake over still another flow in a spectacular waterfall. Outflow from the lake goes under the lava flows in Lost Creek canyon.

McKenzie Pass is on the south flank of Belknap Volcano which, like most basalt volcanoes, is rather broad and flat and doesn't do much for the skyline. Nevertheless, it is an impressive pile of lava flows. The last act at the central crater of Belknap was construction of a cinder cone several hundred feet high putting a pointed cap on an otherwise rather flat volcano. Since then a series of eruptions have poured lava flows from several subsidiary vents on the sides of the volcano.

At McKenzie Pass there is a little lookout building called the Dee Wright Observatory which is made of basalt blocks, and blends so perfectly into the landscape that it is almost invisible. It is an excellent place to look around at what seems like a sea of basalt lava flows so recently erupted that nothing grows on them. They seem perfectly fresh.

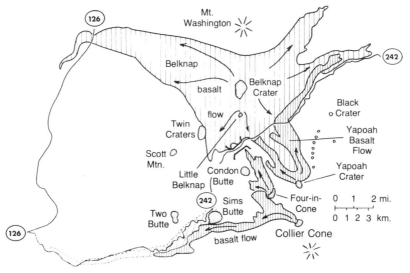

Map showing recently active volcanic vents and lava flows around McKenzie Pass.

Although it all looks at first like a jaggedly uniform sea of rubbly basalt, there are actually several different lava flows at McKenzie Pass. The oldest flows, those northwest of the highway with a scattering of scrawny trees on them, came from the main crater of Belknap volcano and were among its last efforts. The much younger flows northwest of the road, the ones that still look so fresh they might have erupted the night before, came from South Belknap and Little Belknap, two inconspicuous little warts about a mile northwest of the road. It seems incredible that such large lava flows could come from such small volcanoes but there is no doubt that they did. Molten basalt can be fluid enough to spread over a large area without piling up around the vent — that is why basalt volcanoes tend to be so flat.

A straggle of trees clings to life in the rubbly surface of a fresh lava flow at McKenzie Pass. The glacier-carved spire of Three Fingered Jack rises in the background.

Dee Wright Observatory is on a flow erupted from Yapoah Cone which is about 3 miles southeast of the highway and not part of Belknap. A short hike along the skyline trail southeast from Dee Wright Observatory leads to Yapoah Cone and several other very recent small volcanoes. It is fairly easy to see from the highway that the flow from Yapoah Cone laps onto those from the Belknap Craters and is therefore the youngest of the group.

Figuring out the exact age of a young lava flow is hard to do and the results are always surprising because flows usually turn out to be older than they look. The best way is to find charred wood in the flow and determine its age by the radiocarbon method. This would be simple enough if charred wood were common in lava flows but trees overwhelmed by basalt usually float to the surface and burn completely without leaving a trace. Fortunately, one of the latest flows from the main Belknap crater does contain charred remains of trees in several places at its western edge where it nearly reaches Oregon highway 126. Radiocarbon dates on these show that the eruption happened sometime around 1600 years ago. Obviously the flows from the satellite Belknap craters must be younger than this and the one from Yapoah cone younger still. There are reasons for believing that the Yapoah flow must be at least 400 years old.

The Yapoah flow poured nearly 5 miles down the eastern slope of the Cascades following the valley of a small stream. Oregon 242 follows its southern edge east of the pass, down into regions of older rocks covered by thick soils and dense forests which hide the rocks quite effectively even though they are thinner than those on the western slopes of the Cascades. There are very few rocks to see between McKenzie Pass and Sisters; even the Yapoah flow is hidden from the road along most of its length although it is only a short distance north through the trees.

The rubbly surface of a fresh lava flow is no place for bicycles. This is the Yapoah flow near McKenzie Pass.

u.s. 26

portland — madras

The route of U.S. 26 between Portland and Madras crosses part of the Willamette Valley, the old volcanic rocks of the western Cascades, and the young volcanic rocks of the high Cascades. Every bedrock outcrop along the way exposes volcanic rocks of one kind or another.

Portland is in the Willamette Valley. Most of the city is built on valley-fill gravels washed into a broad fault-block basin during Pliocene time, roughly between 3 and 10 million years ago.

Clusters of small volcanoes that were active sometime within the last 10 million years surround the Portland Valley. One geologist recently counted a total of 74 volcanoes within the Portland area, more than a third of them in the hills just south of U.S. 26, between a point several miles west of Gresham and Sandy. His count may well be a bit conservative because some of the older cones are now so eroded that it would be easy to overlook a few of them.

Most of the small volcanoes in the Portland area are basalt cinder cones. These are the little fellows that start erupting off in a field somewhere, cough chunks of bubbly basalt magma out of their vent until they build a cone of cinders a few hundred feet high, and then finish the eruption by producing one or two lava flows that emerge from the base of the cone. Once they quit they never erupt again. Each new eruption produces a new cinder cone. Some of the cinder cones in the Portland area are still fairly fresh so it is conceivable that continued activity may produce new ones in the future.

A few of the volcanoes in the Portland area are broad domes of basalt, the kind that geologists call "shield volcanoes." These form as repeated basalt eruptions from the same vent build a pile of lava

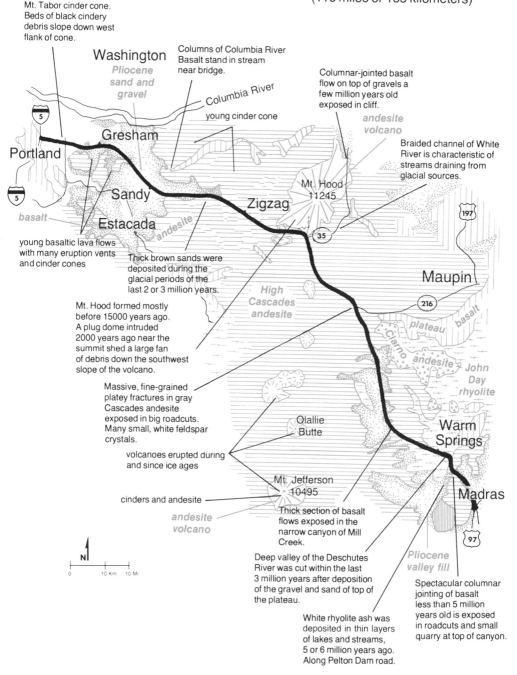

Mt. Tabor cinder cone.
Beds of black cindery
debris slope down west
flank of cone.

Washington
*Pliocene
sand and
gravel*

Columns of Columbia River
Basalt stand in stream
near bridge.

Columnar-jointed basalt
flow on top of gravels a
few million years old
exposed in cliff.

Columbia River

young cinder cone

*andesite
volcano*

Gresham

Portland

Braided channel of White
River is characteristic of
streams draining from
glacial sources.

Mt. Hood
11245

Sandy

Zigzag

35

197

basalt

Estacada

andesite

Maupin

young basaltic lava flows
with many eruption vents
and cinder cones

Thick brown sands were
deposited during the
glacial periods of the
last 2 or 3 million years.

*High
Cascades
andesite*

216

plateau

basalt

Mt. Hood formed mostly
before 15000 years ago.
A plug dome intruded
2000 years ago near the
summit shed a large fan
of debris down the southwest
slope of the volcano.

Clarno

andesite

*John
Day
rhyolite*

Massive, fine-grained
platey fractures in gray
Cascades andesite
exposed in big roadcuts.
Many small, white feldspar
crystals.

Olallie
Butte

**Warm
Springs**

volcanoes erupted during
and since ice ages

N

0   10 Km   10 Mi

Mt. Jefferson
10495

Madras

cinders and andesite

97

*andesite
volcano*

Thick section of basalt
flows exposed in the
narrow canyon of Mill
Creek.

*Pliocene
valley fill*

Deep valley of the Deschutes
River was cut within the last
3 million years after deposition
of the gravel and sand of top of
the plateau.

Spectacular columnar
jointing of basalt
less than 5 million
years old is exposed
in roadcuts and small
quarry at top of canyon.

White rhyolite ash was
deposited in thin layers
of lakes and streams,
5 or 6 million years ago.
Along Pelton Dam road.

flows which tends to be broad and flat because basalt is such fluid magma. Shield volcanoes are never very conspicuous; the high Cascades are full of big ones that people never notice.

Most of the route between the eastern suburbs of Portland and Rhododendron is actually in the western Cascades although there is no way to know that by looking at the rocks along the road. Most of the bedrock in this area is covered by blankets of gravelly sediment washed over it from the higher mountains. Outcrops are hard to find and the road doesn't cross many.

A few miles east of Rhododendron the highway crosses the crest of the high Cascades passing just south of Mount Hood, Oregon's most spectacular volcano. Mount Hood is a big andesite volcano; its great height and steep profile are typical of such volcanoes which consist of a complex mess of lava flows, ash deposits, and mudflows.

Mount Hood hasn't erupted for a long time, probably not for some thousands of years. Big glaciers have carved deep gouges into its flanks and the volcano has done nothing for a long time to avenge the insult by repairing the damage and melting the ice. One of the last acts of Mount Hood was to squeeze up a nearly solid plug of magma near the peak — a plug dome. This kind of activity often signals the end of activity in a volcano. So it is quite possible that Mount Hood is finished and may never erupt again. But it isn't possible to be sure about that. The status of big volcanoes is difficult to determine and several have erupted magnificently after people had declared them dead. So it isn't wise to worry very much about the prospects of an eruption on Mount Hood but neither is it safe to assume that the volcano is dead.

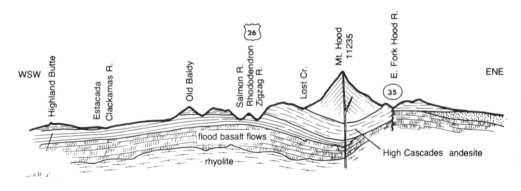

Section through Mount Hood.

The Crooked River Gorge cut through lava flows. On U.S. 97 near Redmond.

If Mount Hood should erupt, it would most likely blow ash and steam and produce a few lava flows. Most of the ash would drift east on the prevailing winds and doubtless cause widespread consternation by causing the rain to fall as mud. A few small earthquakes would give the area a mild shaking and the volcano would make occasional loud rumbling noises. The dark plume of ash and steam drifting east from the summit would make a spectacular sight if the weather cleared up. Any lava flows would stay on the flanks of the mountain without posing much hazard to life or property.

Long shrinkage columns in an exposure of basalt about 7 miles northwest of Madras.

Mudflows are often the most dangerous and destructive effects of an eruption; people tend to underestimate their potential because the name doesn't suggest anything very fierce. Volcanic ash mixes with water falling as rain or coming from melting snow and ice to make thick mud which pours down the slopes of the volcano, picking up boulders along the way. Such slumgullions are capable of travelling at highway speeds and sometimes go 50 miles or more before they finally stop after having buried everything in their path.

Mount Hood is not at all likely to disembowel itself in the sort of grand finale that turned Mount Mazama into Crater Lake. That kind of thing takes rhyolite magma which doesn't seem to exist in Mount Hood or anywhere nearby. No doubt its absence is due to the fact that this part of the Cascades rests on oceanic crust so there is nothing below the surface that could melt into rhyolite. Most of the rhyolite in the Oregon Cascades is south of the Eugene-Bend area where it probably formed by melting Klamath rocks buried beneath the younger volcanics.

The route between the Mount Hood area and Madras crosses a broad apron of basalt lava flows and mudflow deposits which forms the eastern flank of the high Cascades. All the rocks formed within the last few million years. The basalt flows belong to a series of broad and most inconspicuous shield volcanoes which make the foundations of the high Cascades.

The highway crosses the Deschutes River just north of Pelton Dam between Warm Springs and Madras. The river has cut its canyon through the basalt surface, which is quite thin in this area, into the pale volcanic ash of the John Day Formation beneath. These deposits erupted from the southern part of the western Cascades in late Oligocene and early Miocene time, roughly 20 or 30 million years ago, and covered a large area of central Oregon.

The curving pattern in some of these layers of rhyolitic ash shows that they were reworked by running water before finally coming to rest. John Day formation near Pelton Dam.

oregon 58

eugene — junction u.s. 97

Between Eugene and its junction with U.S. 97, Oregon 58 angles across both the western and high Cascades. Every rock along the way is volcanic.

Springfield is at the eastern edge of the Willamette Valley where it meets the western Cascades which continue eastward along the line of the highway to the area about 5 miles east of McCredie Springs. All of the rocks along this part of the route are between 15 and 30 million years old. Some of them are very dark andesites and basalts, others are lighter-colored rhyolites. All erupted during the same long period of volcanic activity but the lighter-colored rocks are generally somewhat older than the dark ones. As everywhere in the western Cascades, soils in this area are so thick and forests so dense that very few rocks of any color are visible except in occasional roadcuts and streambanks. It is frustrating country for geologists.

Right where it passes the dam that impounds Lookout Point Reservoir, the road skirts the north edge of an irregular patch of coarse-

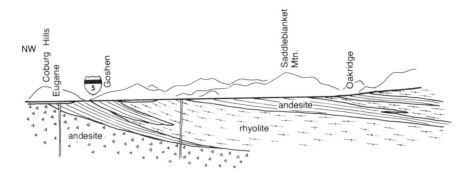

Section along the line of highway 58 from Eugene to U.S. 97.

146

grained granitic rock about a mile in diameter. This is a plug of magma that pushed up into the volcanic rocks and cooled slowly within them without erupting to the surface. The only difference between this kind of rock and the andesitic volcanic rocks surrounding it is the fact that it cooled slowly enough to develop large crystal grains; chemically they are the same.

Along the eastern half of the route, between the area about 5 miles east of McCredie Springs and the junction with U.S. 97, the road crosses the high Cascades through Willamette Pass. Rocks along this part of the route are very dark andesites or basalts and they all look pretty much alike. Although these rocks are fairly young, none of those near the road erupted since the last ice age so they are well covered by soil and trees and hardly more visible than the older volcanic rocks in the western Cascades. Fortunately, there are a number of excellent roadcuts west of Willamette Pass. The Willamette River traces a remarkably straight course southeast of Eugene because it is following a large fault that offsets the southern part of the western Cascades to the northwest.

Numerous large glaciers flourished along the crest of the high Cascades during the last ice age and we owe most of the beautiful scenery near Willamette Pass to their efforts. Some of the glaciers in this area must have been thousands of feet thick because they gouged deep holes in their valleys that now hold large and exceptionally beautiful lakes. The road follows the north side of one of these, Odell Lake, for several miles.

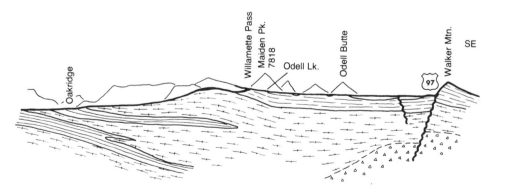

Section along the line of Oregon 58.

58
EUGENE — JCT. U.S. 97
(92 miles or 148 kilometers)

138
ROSEBURG — JCT. U.S. 97
(101 miles or 164 kilometers)

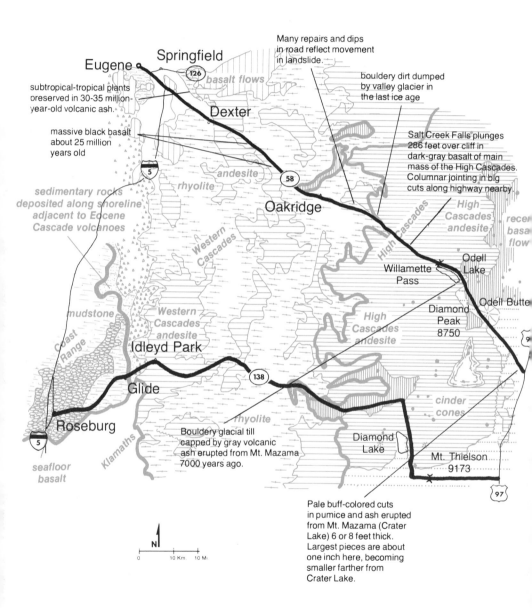

Many repairs and dips
in road reflect movement
in landslide.

bouldery dirt dumped
by valley glacier in
the last ice age

Salt Creek Falls plunges
286 feet over cliff in
dark-gray basalt of main
mass of the High Cascades.
Columnar jointing in big
cuts along highway nearby.

subtropical-tropical plants
preserved in 30-35 million-
year-old volcanic ash.

massive black basalt
about 25 million
years old

sedimentary rocks
deposited along shoreline
adjacent to Eocene
Cascade volcanoes

High
Cascades
andesite

recei
basa
flow

Western
Cascades

mudstone

Western
Cascades
andesite

High
Cascades
andesite

Odell
Lake

Willamette
Pass

Diamond
Peak
8750

Odell Butte

Coast
Range

Idleyd Park

Glide

Roseburg

seafloor
basalt

Klamaths

rhyolite

Bouldery glacial till
capped by gray volcanic
ash erupted from Mt. Mazama
7000 years ago.

Diamond
Lake

cinder
cones

Mt. Thielson
9173

Pale buff-colored cuts
in pumice and ash erupted
from Mt. Mazama (Crater
Lake) 6 or 8 feet thick.
Largest pieces are about
one inch here, becoming
smaller farther from
Crater Lake.

Eugene

Springfield

126

basalt flows

Dexter

58

Oakridge

andesite

rhyolite

N

0   10 Km.   10 Mi.

East of Willamette Pass, the country begins to get a bit drier and the forest thinner because most of the rain and snow fall on the western slope and crest of the Cascades. Between Crescent Lake and U.S. 97, scrawny trees grow in a thick deposit of pea-sized bits of white pumice which drains very rapidly and keeps them from getting enough water. This is the northern edge of the huge blanket of pumice and volcanic ash that Mt. Mazama laid down when it disemboweled itself a little more than 7000 years ago in the eruption that created Crater Lake.

Close view of frothy pumice erupted from Mount Mazama during its big extravaganza. The biggest chunks in this picture are about the size of grapefruit.

oregon 138

roseburg — junction u.s. 97

Oregon 138 follows a route through beautiful and wild country all the way between Roseburg and the junction with U.S. 97. It is an unusually interesting drive through an extraordinary variety of fascinating rocks which would be easier to see if the thick soils and somber forests along much of the route didn't hide them quite so effectively.

Roseburg is near the eastern edge of the Coast Range just north of where it abuts the older rocks of the Klamaths. The hills around town are eroded in basalt lava flows that were part of the Pacific Ocean floor back in Eocene time, perhaps 50 million years ago. Between the point where the highway leaves the old seafloor basalts about 7 miles east of Roseburg and the area about a mile west of Idleyd Park, it crosses a series of dirty-looking sandstones and mudstones deposited on the seafloor about the same time the basalt lava flows were erupting.

For approximately 35 miles between the area just west of Idleyd Park and that in the vicinity of Clearwater, Oregon 138 crosses volcanic rocks of the western Cascades erupted between 35 and 20 million years ago. Erosion has since carved them into the rugged ridge and ravine landscape we see today in which almost no trace of the original volcanic mountains survives. Except for occasional basalt lava flows, most of the rocks in this part of the western Cascades are andesites and even lighter-colored rocks.

The short stretch of road between Steamboat and Eagle Rock campground passes through volcanic rocks bleached and altered almost beyond recognition by hot water and steam. No doubt this has something to do with the fact that the road also crosses a small mass of granite just west of Eagle Rock campground. This one is about a mile in diameter and the road goes right through it past exposures of coarsely granular gray rock.

Most of the small bodies of granite in the western Cascades are surrounded by at least 15 or 20 square miles of volcanic rocks altered to some degree by the action of hot water and steam. A plug of hot granite magma stabbed into the inside of a volcano might easily take many thousands of years to cool and in the meantime its heat will drive a circulation of hot water that cooks the insides of the volcano as though they were a Christmas pudding. The hot water and steam dissolve some of the original minerals, change others, and bring in some new ones. If a significant mining industry ever develops in these mountains, it will most likely center in an area of altered rock surrounding a granite plug.

Between the south end of Diamond Lake and U.S. 97, the highway crosses the field of pumice laid down a little more than 6000 years ago when old Mount Mazama blew out its insides to create Crater Lake. Trees grow poorly on the rubbly pumice field because water quickly soaks down beyond the reach of their roots leaving them gasping for moisture. And the pumice is so young and fresh that weathering has hardly begun to reduce it to soil so the trees are growing on what is essentially a rubble of freshly broken rock.

Pumice is a frozen foam of rhyolite glass so full of air bubbles that most chunks of it will float nicely on water. It is valuable for making lightweight concrete aggregate with excellent structural, insulating, and soundproofing qualities. The only reason pumice isn't mined more extensively in this part of Oregon is that there is so little local demand for building materials.

Diamond Lake is a gorgeous thing. A big glacier gouged its basin during the last ice age leaving as its legacy one of the finest lakes in the Cascades.

Pebbles turning in the grip of eddies drilled these potholes in the bed of the Rogue River at Union Creek. The rock is basalt.

oregon 62

medford — junction u.s. 97

The route from Medford to U.S. 97 loops north through the western Cascades and then crosses the high Cascades through the area affected by the last great eruption of Mount Mazama that created Crater Lake nearly 7000 years ago. All the rocks along the way are volcanic; those in the western Cascades erupted about 25 to 35 million years ago, most of those in the high Cascades date from the Mazama eruption.

The short section of road between Medford and Eagle Point crosses soft sedimentary rocks, mostly sandstone, that accumulated on land some 50 million or so years ago. These cover older sedimentary rocks,

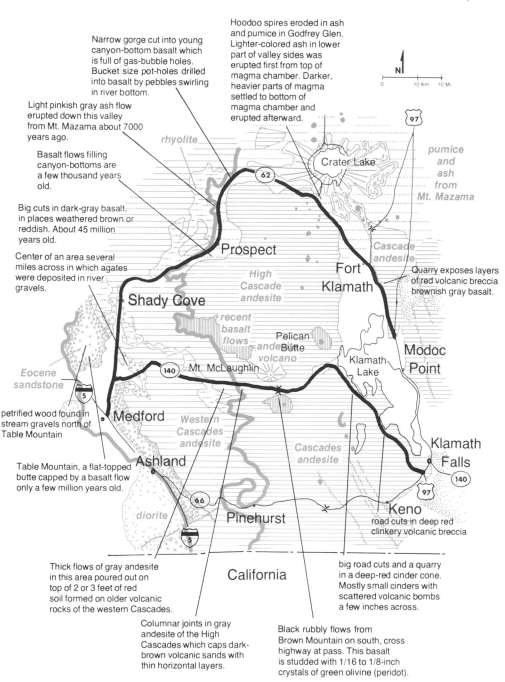

Narrow gorge cut into young canyon-bottom basalt which is full of gas-bubble holes. Bucket size pot-holes drilled into basalt by pebbles swirling in river bottom.

Hoodoo spires eroded in ash and pumice in Godfrey Glen. Lighter-colored ash in lower part of valley sides was erupted first from top of magma chamber. Darker, heavier parts of magma settled to bottom of magma chamber and erupted afterward.

Light pinkish gray ash flow erupted down this valley from Mt. Mazama about 7000 years ago.

Basalt flows filling canyon-bottoms are a few thousand years old.

Big cuts in dark-gray basalt, in places weathered brown or reddish. About 45 million years old.

Center of an area several miles across in which agates were deposited in river gravels.

Quarry exposes layers of red volcanic breccia brownish gray basalt.

N

0   10 Km   10 Mi

rhyolite

Crater Lake

pumice and ash from Mt. Mazama

62

Prospect

Cascade andesite

High Cascade andesite

Fort Klamath

Shady Cove

recent basalt flows

Pelican andeButte volcano

Eocene sandstone

Mt. McLaughlin

Klamath Lake

Modoc Point

140

petrified wood found in stream gravels north of Table Mountain

Medford

Western Cascades andesite

Cascades andesite

Klamath Falls

Table Mountain, a flat-topped butte capped by a basalt flow only a few million years old.

Ashland

140

97

66

diorite

Pinehurst

Keno

road cuts in deep red clinkery volcanic breccia

5

Thick flows of gray andesite in this area poured out on top of 2 or 3 feet of red soil formed on older volcanic rocks of the western Cascades.

California

big road cuts and a quarry in a deep-red cinder cone. Mostly small cinders with scattered volcanic bombs a few inches across.

Columnar joints in gray andesite of the High Cascades which caps dark-brown volcanic sands with thin horizontal layers.

Black rubbly flows from Brown Mountain on south, cross highway at pass. This basalt is studded with 1/16 to 1/8-inch crystals of green olivine (peridot).

buried in this area, that formed along the shores of a sea that existed east of the Klamath Mountains in late Cretaceous time, about 75 million years ago.

The road between Eagle Point and McLeod crosses the older volcanic rocks of the western Cascades. Those along this stretch of the road are mostly dark gray andesites and deposits of lighter-colored volcanic ash.

Almost all of the long route between McLeod and Fort Klamath crosses ash-flow deposits erupted from Mount Mazama when it turned itself inside out nearly 7000 years ago. The surface of these deposits is mostly pumice, a very pale rock that consists essentially of rhyolitic volcanic glass puffed up into a foam of tiny bubbles. Big chunks of pumice make lightweight garden rocks that can be rearranged like so much furniture.

The ash-flow deposits around the remains of Mount Mazama fill old valleys that existed before the eruption. The route between McLeod and Crater Lake National Park follows one of those valleys and that between the park and Fort Klamath another.

Just north of Fort Klamath the road crosses a steep escarpment and drops down off the surface of the recent volcanics onto the floor of a broad fault block valley dotted with marshes. Klamath Lake floods it a bit farther south. Deposits of volcanic ash laid down in lake water lie beneath its floor.

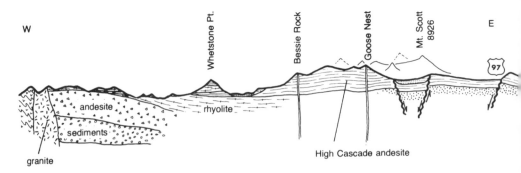

Section south of Crater Lake showing the volcanic rocks of the Cascades piled onto the much older rocks of the eastern flank of the Klamaths.

crater lake national park

Crater Lake is a gem created in catastrophe. A towering snowcapped volcanic cone at least as magnificent as Shasta stood here until about 7000 years ago when it destroyed itself in a cataclysmic eruption that devastated hundreds of square miles of the surrounding countryside and sent a blanket of white ash sifting down over a large part of the northwest. The sky must have been dark for days. All that remains of Mount Mazama is a stump with Crater Lake in its center.

It is possible to reconstruct the approximate shape of Mount Mazama by extending the profiles of its remaining lower slopes upward at angles like those of other big andesite cones. They outline a slightly lopsided mountain which had its summit above the south rim of Crater Lake and was somewhere between 10,000 and 12,000 feet high depending upon how much its crater may have lopped off the top.

We can be quite sure that Mount Mazama had big glaciers on it because the remnants of valleys on its surviving flanks show the clearly written signature of ice. They are gouged out into the broad cross-section typical of heavily glaciated valleys and contain bedrock surfaces scraped and polished by glacial ice. The glaciers must have shrivelled greatly by the time of the big eruption because the last ice age ended about 10,000 years ago. But small glaciers surely remained clinging to the higher slopes just as they do today on Mount Hood.

It seems likely that Mount Mazama may have been fairly quiet for a long time before its big eruption. Big glaciers had carved its sides and the local Indians probably considered it to be an extinct volcano if they worried themselves about such things. But during that long period of quiet, a large batch of rhyolite was slowly cooking beneath the volcano, absorbing water that would later blow out the vent as live steam. Quiet volcanoes aren't always dead.

Most of the batches of rhyolite that cook up beneath big volcanoes probably cool beneath the surface without erupting. They become granitic intrusions. Large masses of granite are certainly far more numerous in the deeply eroded roots of old volcanic chains than massive eruptions of rhyolite are in the active chains. But accidents do happen and sometimes the magma blows out through a volcano instead of crystallizing into granite.

The eruption may have begun fairly innocently, perhaps with a big plume of steam and ash blowing out the top of Mount Mazama and drifting gracefully downwind. But the escaping steam and ash must have cleaned out the volcanic plumbing and relieved pressure on the steaming mass of rhyolite beneath, permitting it to blow off more steam which couldn't escape as rapdily as it formed because molten rhyolite has a consistency about like cold peanut butter. So most of the steam was trapped in the magma and foamed it up into pumice and ash which boiled out of the volcano and down its slopes as though it were oatmeal overflowing a pan.

The red-hot mixture of steam, pumice, and ash that boiled out of the volcano and flowed down its slopes and across the surrounding countryside was so hot that the rock fragments welded themselves together as soon as the mass stopped moving. We now see them as the huge welded ash flows that surround Mount Mazama today. They are covered by a deep rubble of loose and unwelded ash and pumice which forms a smooth surface and supports a very sparse growth of stunted trees.

Not all the ash poured down the sides of the volcano in the big welded ash flows. Some of it blew high into the sky and drifted on the wind, finally settling to become an ash fall that covered much of the northwest. Most of the ash blew northeast and deposits of it form part of the soil in places as far away as Fort Benton, Montana. Unlike ash flows, ash falls are never welded and always make soft deposits easily reworked by wind and water.

As the magma erupted from beneath Mount Mazama, the volcano lost its foundations and sank into the ground, leaving a hole where there had been a mountain. Years ago many people believed that the volcano had blown itself apart in a great explosion. But such a blast would have strewn shattered fragments of the volcano all over the surrounding countryside. As it happens, only a small part of the debris blanket surrounding Mount Mazama consists of chunks of old volcanic rock, most of it is freshly-erupted pumice and ash. So Mount

Mazama could not have blown itself apart and must have simply subsided as the eruption withdrew magma from beneath it. The eruption surely included some mighty explosions but they blew fresh pumice, not old volcanic rock, all over the landscape.

Geologists estimate that approximately 10 to 12 cubic miles of rock erupted from Mount Mazama during its last great outburst and that approximately 15 to 17 cubic miles of mountain disappeared. Both estimates are difficult to make and obviously subject to all sorts of uncertainties but the difference between them is still too large to be ignored. No one is quite sure how to account for it.

The eruption continued after the mountain had collapsed, starting two new cones on the floor of the crater. One of these is high enough to rise above the surface of Crater Lake as Wizard Island, the other is submerged. There is no way to know whether activity will continue with more eruptions but neither is there any reason to assume that Mount Mazama is extinct. Other volcanoes have collapsed into similar caldera craters and then rebuilt themselves and continued with a long history of eruptions. Mount Lassen is a good example and Mount Vesuvius is another.

The water in Crater Lake comes entirely from rain and melting snow in a watershed that is extremely small for such a big lake. The annual input and output of water is so small that Crater Lake doesn't rapidly flush itself. Therefore, it would be an easy lake to pollute and a difficult one to clean up. It is a delicate and vulnerable gem that must be treated with care.

View along the rim of Crater Lake from Hellman Pk. lookout to Llao Rock.

oregon 140

medford — klamath falls

The drive from Medford to Klamath Falls starts at the eastern edge of the Klamaths and passes all the way through both the old western Cascades and the active high Cascades. It is an interesting route.

Medford is at the eastern edge of the Klamaths, sitting on sedimentary rocks deposited on land during Eocene time, about 50 million years ago. These and some older sedimentary rocks deposited in sea water during late Cretaceous time, about 75 million years ago, lap onto the eastern edge of the Klamaths all the way down to their southern end near Redding, California. Those older sediments deposited in sea water are significant, the fact that they lap onto the edge of the Klamaths can only mean that there was an open sea east of the Klamaths during late Cretaceous time. They are the evidence that tells us that the Klamaths were at least a peninsula, if not an island, back then.

Between Eagle Point and the area about 3 miles west of Fish Lake, the road crosses the older volcanic rocks of the western Cascades as it passes through sunny slopes dotted with oak trees at low elevation on the west and through dark evergreen forests higher in the mountains to the east. There are numerous roadcuts, all of them in andesites, to compensate for the scarcity of good natural exposures. The wild purplish and greenish shades of many of the rocks suggests that they are altered by steam and hot water. As usual, most of the andesites are masses of broken fragments, called "agglomerates," rather than solid lava flows.

All of the rocks between the area a few miles west of Fish Lake and Klamath Falls are young volcanic rocks of the active high Cascades. Fish Lake is right between Mount McLoughlin, north of the highway, and Brown Mountain, south of the highway, both of which are so fresh and unmarked by erosion that they must surely be active volcanoes. There are lava flows beside the road at the top of the pass between them which are old enough to have some trees growing on them but still young enough to make ragged heaps of rubbly lava in the sparse woods.

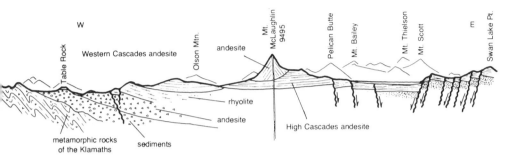

Section near the line of Oregon 140 between the area west of Medford and Klamath Falls.

Most of the country on the east slope of the Cascades is dry enough that there are numerous good outcrops of dark volcanic rocks, the black ones are basalt and all those that aren't quite that dark are andesite. Some of the weathered rocks are stained bright red by iron oxides.

Klamath Lake fills a basin created when a large block of the crust dropped along faults during the latter part of the last million years. The fault lines are very easy to see because they make long, straight scarps on both sides of the lake.

A few trees are gaining a foothold in the rubbly basalt surfaces where Oregon 140 crosses the crest of the Cascades.

159

This stockade of weathered columns is an andesite flow resting on dark volcanic ash beside the road about 6 miles west of the pass on Oregon 140.

the volcanic plateaus

An enormous region extending from north-central Washington to northeastern California and including most of Oregon east of the Cascades is covered by basalt lava flows erupted during Miocene time, between 20 and 15 million years ago. Generations of geologists regarded this vast region as a single geologic province; they called it the "Columbia lava plateau" and left it at that. No one looked very carefully at the basalt or thought very deeply about what it might mean or what older rocks it might cover.

All that is changed now. During the last two decades a number of geologists have begun to examine the region rather closely and find that the geologic picture is both complex and fascinating. The lava plateau contains subtly different kinds of basalt erupted from different places at different times and there are some other kinds of rocks too. So now instead of a big, simple geologic province we have a puzzle in many parts and no one can yet put all the pieces together and make the picture whole. The region is much more interesting now that we have begun to learn something about it instead of just giving it a name.

The oldest rocks in the plateau country are the Blue and Wallowa Mountains which stand like islands above the flood of basalt. When they were new, they were part of a chain of coastal mountains that embraced a bay of the ocean that has since become most of central and northern Oregon. The best window into that old bay is in the areas around Mitchell and south of Dayville where sedimentary rocks that accumulated in it are now exposed at the surface. They are dirty sandstones and mudstones, many of them full of volcanic ash, which were deposited during much of Triassic, Jurassic, and Cretaceous time, approximately between 200 and 80 million years ago.

These rocks were deposited on a slab of moving seafloor that was crowding them back against the continent so we find them crumpled into folds. Their content of volcanic ash suggests that the bay was fringed by volcanoes, perhaps a chain of volcanic islands.

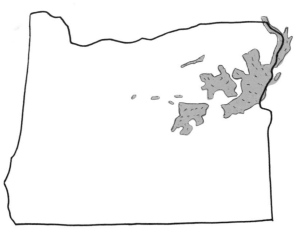

Map showing distribution of Triassic, Jurassic, and Cretaceous sedimentary rocks in north-central Oregon. They must be much more widespread beneath the basalt.

By Eocene time, about 50 million years ago, the seafloor was sinking along a line well west of the Blue-Wallowa-Klamath arc, and the fringing chain of volcanoes followed the line of the present western Cascades about as far north as Eugene where it swung northeastward toward Pendleton. The ocean bay still existed but had become somewhat smaller. It was finally cut off mostly from the ocean and converted into an inland sea about 35 million years ago when the line of volcanic islands shifted to the present north-south trend of the western Cascades.

Part of that late Eocene volcanic chain still stands high above the surrounding sea of younger basalts in the Ochoco Mountains, a complex pile of dark lava flows, mudflow deposits, and ash beds locally interlayered with sedimentary rocks. Geologists lump the whole dark-colored mess together and call it the Clarno Formation. Rocks in the Ochocos are the same age and closely resemble the Colestin Formation in the westernmost western Cascades south of Eugene except that

162

they don't contain any sediments deposited in seawater. Presumably those must exist somewhere north of the Ochocos where everything is buried under basalt lava flows.

Some of the ash beds and sedimentary layers in the Clarno Formation contain magnificent deposits of fossil leaves, mostly rather tropical-looking leaves, the sort of thing you might expect to find growing today in Central America. And we also find red soils buried in the Clarno Formation like those that form today in warmly humid regions.

After the Clarno eruptions ended, there was a long period of volcanic quiet during which erosion bit deeply into the volcanic rocks, carving them into a rugged landscape of hills and ravines. It was during this quiet time that the line of seafloor sinking shifted from the curving path it had been following around the margin of the old ocean bay to its present straight course offshore from the modern coastline. The old volcanoes quit erupting when the seafloor stopped sinking along the old line. It took a few million years for the slab of seafloor sinking along the new line to get deep enough to begin melting and start new volcanoes.

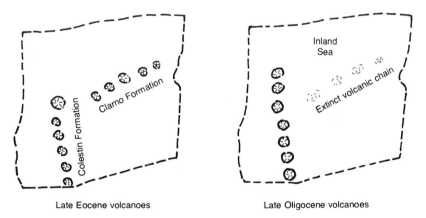

Late Eocene volcanoes   Late Oligocene volcanoes

A shift in the line of volcanic activity created an inland sea about 35 million years ago.

When the new volcanoes did start erupting in the western Cascades between 25 and 30 million years ago, they sent enormous clouds of light-colored ash drifting over central

Oregon. It accumulated to depths of more than 1000 feet in some places to become what we call the John Day Formation. Like most deposits of light-colored volcanic ash, the John Day Formation is quite colorful, being reddish in its lower part and grading upward through a greenish middle section to pale yellowish brown rocks near the top. The middle and upper parts of the formation contain numerous fossil leaves and bones which people have been collecting for more than a century. They tell us of a moist, mild climate and a landscape clothed in lush vegetation.

Most of the John Day Formation is rather soft rock composed of ash that drifted in on the wind. But it includes ledges of very hard rock composed of volcanic ash welded into a solid mass. Welded ash forms when a volcano erupts a cloud of steam mixed with shreds of molten rock which surges across the countryside at speeds well in excess of any highway speed limit. When the cloud finally settles, the fragments of molten rock weld themselves together into a solid mass — they are that hot. Such eruptions are explosively violent and every bit as destructive as they sound; everything in the path of those glowing hot clouds of molten rock is instantly incinerated. Several such eruptions have been observed in historic time but none of those produced welded ash beds that went more than a few miles from the vent or covered more than a few square miles. The eruptions that put the welded-ash beds in the John Day Formation must have come from the western Cascades, probably somewhere south of Salem, and the welded-ash ledges extend eastward at least as far as Dayville and cover thousands of square miles. No volcanic eruption on a scale even remotely comparable to that has ever been witnessed in historic time. Those innocent-looking ledges of hard rock in the John Day Formation tell a story of natural violence on a scale never recorded in human history.

Then, in early Miocene time, about 25 million years ago, the western Cascades quit erupting and the volcanic action shifted first to central and then to eastern Oregon. This is an unusual event, so something strange must have happened to cause it. As we showed in our first chapter, we suspect that the descending slab of seafloor may have broken off not far below the surface and sunk more rapidly into the earth's interior. This would certainly have put a stop to volcanic activity in the western Cascades and it is easy to imagine that material flowing westward beneath the crust to fill in behind the broken slab might have pulled the crust westward, opening cracks that would trend in a north-south direction.

There is no doubt that basalt comes from the dark rocks of the earth's interior, from beneath the crust. Those rocks are normally solid, or nearly so, even though they are hot enough that they would melt were they not under such tremendous pressure. If the pressure on them is relieved, as happens if cracks open in the crust above them, they begin to melt and the liquid they produce is basalt.

Rocks beneath the crust are not basalt but something quite different called periodotite. It seems strange at first that basalt should have a composition different from the rocks that melt to produce it but the phenomenon is actually quite familiar in other situations. For example, we have all watched a small child wave a popsicle around in the sun until it begins to drip and tragedy seems immiment and then, with loud slurping noises, suck all the juice out leaving white ice on the stick. Everyone in the nursery school set knows that popsicle juice has a lower melting point than ice. Similarly, basalt has a lower melting point than the rocks in the earth's interior so it is the liquid they produce when they partially melt.

Most of the Miocene volcanic activity in central and eastern Oregon involved eruption of enormous floods of basalt, some of which covered thousands of square miles with single lava flows having volumes measurable in hundreds of cubic miles. They are well named; they really were floods of basalt lava. No such overwhelming eruptions of basalt have happened anywhere in the world during historic time so we have no eyewitness accounts to help us picture what they were like.

Molten basalt has a consistency about like cold molasses which makes it very fluid and runny by magmatic standards. The spread of a basalt flow on level ground — and much of the basalt plateau is nearly level — is limited mainly by the clinkery crust that forms on the cooling surface and then tips over the front of the flow making a sort of natural dam. As long as the flow is more than about 50 feet thick, the pressure of the molten basalt will generally be enough to overwhelm the clinker dam and the lava will continue to spread. When the flow gets less than 50 feet thick, it begins to have trouble bursting through its clinkery front and will stop before going much farther. Some of the flood basalt flows are hundreds of feet thick and traceable for a distance of more than 200 miles.

Geologists have had a hard time coping with the volcanic plateau because the basalt flows are very large, tend to look very much alike, and seemed for a long time to have no obvious source. These are all difficult problems and none are completely solved. But there are different kinds of basalt and anyone who cares to study them closely enough can learn to tell them apart. Then it becomes possible to subdivide the plateau into regions according to their types of basalt and to determine the relative ages of the different kinds by observing which lie on top of others. And it is also possible to identify the sources of the flows in basalt dikes which are simply the fissure from which the magma poured and now filled with solid basalt. By matching the kinds of basalt in the dikes with those in the flows, it is possible to sort the volcanic plateau into a mosaic of coalescing giant volcanoes.

The oldest flood basalt eruptions seem to have come from a swarm of large basalt dikes which run approximately north-south in the region between Monument and Dayville in north-central Oregon. These dikes, and the lava flows that came from them, are composed of granular basalt flecked with crystals of a pale green mineral called olivine which easily weathers to rusty specks. When this basalt begins to break down into soil, it forms rounded chunks of rotten rock which are very distinctive. It is called the Picture Gorge basalt because it is so nicely exposed in the walls of Picture Gorge.

The Picture Gorge volcano, if you choose to call it that, covered a large area almost entirely south of Battle Mountain ridge which extended east as far as the mountains in Idaho. There is no way of knowing how far west or south the Picture

Gorge flows may have gone because they disappear under younger rocks in both those directions.

After the Picture Gorge volcano had about finished its career, new volcanoes began erupting in the Strawberry Mountain area southeast of John Day. Some andesite and enormous quantities of rhyolite erupted there, forming what must have been an imposing group of volcanoes. The light-colored rhyolite ash drifted and washed over much of central Oregon burying the Picture Gorge volcano.

Rhyolite is at the opposite end of the volcanic spectrum from basalt; it is very light-colored and rich in silica as opposed to basalt which is black and relatively poor in silica. Continental rocks, such as the deformed sandstones and mudstones that make coastal mountains, are rich in silica and will yield rhyolite if they partially melt. Deeply buried continental rocks are hot enough that they would melt were they not under great pressure, so opening cracks in the crust will cause rhyolite magmas to appear simply by relieving the pressure on silica-rich rocks at depth.

So there is no problem imagining how basalt or rhyolite magmas may appear in places where the earth's crust is stretched. Either or both may form depending upon whether or not thick accumulations of hot, silica-rich rocks exist in the crust. But the andesites in the Strawberry Mountains are awkward. They are intermediate in composition between basalt and rhyolite and such magmas cannot form simply by relieving pressure on the already-hot rocks deep within or below the crust. Those andesites in the Strawberry Mountains don't seem to fit into the picture in any simple way. We can only suggest that they may have formed because, for some reason, the basalt magmas did not rise directly to the surface but lingered on the way long enough to assimulate some silica-rich rocks and convert themselves into andesite. That may well be what happened, but it doesn't explain why it happened.

While the Strawberry Mountain activity continued in north-central Oregon, new basalt flood eruptions began in the northeastern corner of the state. Thousands of large basalt dikes exposed in the high canyon walls in that area seem to have been the feeder conduits for a series of incredible lava flows that covered all of Oregon north of Battle Mountain ridge as well as most of southern and central Washington. Some of the flows poured down the valley of the ancestral Columbia River all the way to the Pacific Ocean and into the northern part of the Willamette Valley. Others flooded eastward against the mountains in western Idaho.

This Grande Ronde volcano, as we prefer to call it because its feeder-dikes are best exposed in Grande Ronde canyon, utterly dwarfed the earlier Picture Gorge volcano and may well be the largest single volcano known on earth. There are larger ones on Mars. Its lava flows, which geologists call the Yakima basalts, break on smooth instead of granular surfaces and lack olivine. They do not weather into rounded cobbles and frequently form prominent ledges marked by rows of crisply outlined vertical columns. No rhyolite erupted from the Grande Ronde volcano, presumably because there are no thick accumulations of silica-rich rocks beneath the northeastern corner of Oregon.

Another volcanic center near the north end of what has since become Steens Mountain covered a large area of southeastern Oregon with a distinctive kind of basalt which often contains flat crystals of white feldspar about the size of postage stamps. The flows must have been unusually fluid because they cover a vast area even though they are relatively thin. There is at least another sequence of lava flows of about the same age in the Modoc Plateau of northeastern California and still another in parts of southeastern Oregon and adjacent Idaho. Volcanoes in southeastern Oregon erupted large amounts of rhyolite as well as basalt which tells us that there are thick accumulations of silica-rich rocks in the crust beneath that part of the state. Presumably these are rocks similar to those exposed at the surface in the Blue Mountains.

We can be sure that many of the flood basalt eruptions were widely spaced in time because we find old soils buried between the flows, soils that developed on an older flow and then were buried by the next one. Some of those soils are very thick and

and must represent lapses of at least hundreds of thousands of years between successive flows in the same place. If we remind ourselves that the flood basalts erupted over a period of perhaps 10 million years, it becomes easy to imagine long intervals between eruptions. So the Miocene was not a period of continuous volcanic uproar even though it did see an incredible amount of activity.

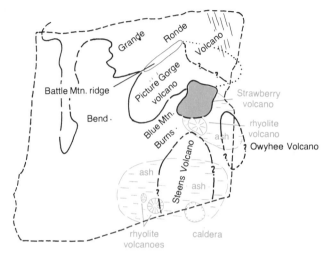

Map showing the patchwork mosaic of large volcanoes that flooded much of Oregon during Miocene time.

The old soils sandwiched between the Miocene lava flows are always easy to see because they are bright red laterites, a type of soil that develops only in wet, tropical climates. No such soils are forming today anywhere very far north of the Tropic of Cancer. They tell us as clearly as anything can that during Miocene time Oregon steamed in a wet, tropical climate and was covered by lush jungles. From time to time one of the great eruptions seared a terrible swath through the forest incinerating everything in its path.

Those huge lava flows frequently dammed streams, creating lakes and swamps which lasted for a while before they finally filled with sediment or drained as erosion slowly lowered their outlets. Much of the sediment that accumulated in them was white volcanic ash, rhyolite, that drifted in on the wind. There is no better way to preserve fossil leaves than to bury them beneath a thin layer of fine, white volcanic ash settling to the bottom of a lake. They look as good as herbarium specimens lovingly dried between sheets of blotting paper. Old lake beds large and small are common in the volcanic plateau and they usually contain numerous leaves beautifully pressed between the thin layers of white volcanic ash. So we have a direct and reasonably complete record of the trees that flourished on those red soils while the flood basalt eruptions slowly built the plateau and they are indeed tropical. The climate must have been hot.

The largest of the flow-dammed lakes fluctuated over an area of several thousand square miles in the Boise-Ontario region of Idaho and eastern Oregon. It must have been overflow from that lake which established the lower course of the Snake River and started carving Hell's Canyon sometime near the end of Miocene time, about 12 million years ago.

All of the Miocene flood basalts in Oregon are peculiarly brown looking; there is always a thin wash of rust over the basic blackness of the rock. Basalts should be black. We suspect that this unusual color tells us once again that these lava flows erupted into a region with a steaming hot and very wet climate

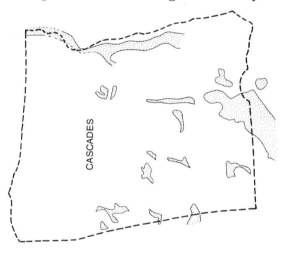

Areas of valley-fill sediment deposited east of the Cascades during Pliocene time when the climate was arid. We suspect that the line of deposits aligned south of the Columbia River may fill the valley of the ancestral Columbia River.

170

and are all at least slightly weathered. One of the first things that happens to basalt when it starts to weather is that certain iron minerals begin to break down, releasing iron oxides which give the rock a rusty stain. The climate dried out and became cooler at the end of Miocene time and all the basalts erupted since then are black.

It must have been just about at the end of Miocene time, about 12 million years ago, that the lush years ended in Oregon and the climate became extremely dry and much cooler. It stayed that way during the next 10 million years, throughout Pliocene time. Dry regions normally have a very high rate of soil erosion because there are so few plants to protect the ground. Neither is there much stream flow to carry the eroded debris away. So the occasional rains wash soil off the hills and into the nearest valley where they dump it. Times of dry climate record themselves in valley-fill deposits which usually contain fossil bones and occasional pieces of petrified wood.

Since the end of Pliocene time, during the last 3 million years of Pleistocene time, the climate of eastern and central Oregon has fluctuated wildly as at least 4 major ice ages came and went. The ice ages were very wet times that were probably somewhat colder than our modern climate but not as extremely cold as most people imagine. Periods between ice ages were dry, perhaps even drier than our modern climate. The coming and going of ice ages was more a matter of switching between wet and dry climates than between cold and warm ones.

The last eruptions from the monstrous Grande Ronde volcano were the final events in the volcanic history of northeastern and north-central Oregon. They happened about 12 million years ago. By that time, nearly all of Oregon east of the Cascades was a vast volcanic plateau covered by basalt lava flows which buried all the older rocks except for the higher parts of the Blue, Wallowa, and Ochoco Mountains. That was just about the time that the climate changed from being hot and wet to cold and extremely dry. Since then the formerly level surface of the lava plateau in northeastern and north-central Oregon has been warped into a few broad folds and broken by a few faults but has otherwise lain almost undisturbed by the earth's internal commotions. About 12 million years of weathering and erosion have created the deep soils and broad valleys that make the fertile countryside and open landscape we see today. Meanwhile, the southern part of the volcanic plateau has continued to be a region of active faulting accompanied by volcanic eruptions. Both continue unabated in that region today.

The boundary between the relatively quiet northern and very active southern part of the modern volcanic plateau is the Brothers fault zone which cuts diagonally southeastward through Oregon from Bend to the vicinity of Jordan Valley. The boundary between the two regions is obvious because everywhere south of the Brothers fault zone the landscape is broken into a dramatic pattern of mountains and broad valleys separated by abrupt fault escarpments. The contrast to the gentler landscape of the northern volcanic plateau is striking.

A first look at a map showing the faults in Oregon south of the Brothers fault zone is a real shock. It looks as though it might be a diagram of breakage in a sheet of glass just run over by a heavy truck. Satellite pictures give the same impression because most of the faults are visible in the landscape. Crustal movement is so recent, indeed still continuing, that erosion has hardly rounded the edges of the fault scarps.

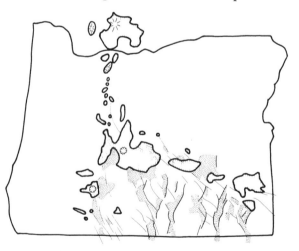

Distribution of younger volcanic rocks and pattern of major faulting in the volcanic plateau. Valley-fill sediments are stippled areas on the map.

Although the fault pattern in southern Oregon looks chaotic at first glance, more careful analysis shows that it is actually simple and orderly. Most of the faults fall into two major groups: one trending nearly northwest and the other slightly east of north. They meet at an angle of about 55°. All available evidence shows that both sets of faults are moving simultaneously so they must be responding to the same stresses and there must be a single explanation for all of them.

So far as is known, all the faults in the group that trend northwest slide sideways with the south side moving northwest. All those in the group trending slightly east of north move vertically with one side rising to form a mountain while the other drops to make a valley. These are the faults that have had the most obvious effect on the landscape but both sets are equally important to the overall scheme.

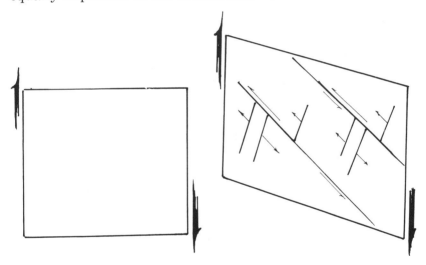

Large arrows show the direction of major crustal movement and small arrows the pattern of movement along faults in southeastern Oregon.

There are various ways of trying to figure out what a pattern of faults may mean. The simplest and most direct is to simulate it with a model. For example, a pasty mixture of cornstarch and water spread like thick cake frosting on a sheet of foam rubber will make a crude model of the earth's brittle crust floating on a plastic interior. Then it is possible to stretch or twist the rubber to simulate movements in the interior and see how the crust fractures in response. Models of this sort make fracture patterns duplicating that in southern Oregon if the west side of the sheet of rubber is pulled north and the east side south so the crust is continuously sheared between them. Other methods of analyzing fracture patterns give the same result. So it appears that the faulting in southern Oregon is due to relative northward movement of the west side of the state.

Faulting and volcanic activity have gone hand in hand in southern Oregon. The pattern of fault movement is stretching the crust and of course this relieves pressure on the very hot rocks at depth permitting them to partially melt. The eruptions have produced rhyolite as well as basalt magmas so we must conclude that thick accumulations of silica-rich rocks are buried beneath the volcanic crust that covers southern Oregon. You can tell what is in a pie by looking at the kind of juice that bubbles up through its crust. Presumably those silica-rich rocks beneath southern Oregon must be the buried part of the Blue Mountains and, if so, we can trace their extent by observing the present distribution of rhyolite on the surface. We find rhyolites throughout southeastern Oregon almost as far west as Klamath Falls but none between there and the Klamath Mountains. So evidently there is a gap about 60 miles wide between the buried western end of the Blue Mountains and the Klamath Mountains.

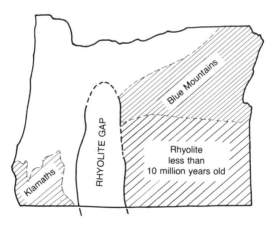

The distribution of rhyolite magmas in Oregon defines the area underlain by older silica-rich rocks like those in the Blue Mountains.

174

The most recent eruptions in the volcanic plateau region have occurred right along the Brothers Fault zone. There are patches of very young volcanic rocks all along the line. The biggest of these is Newberry volcano, just south of Bend, which is usually included in the Cascades but it is unlike any other Cascade volcano, stands about 50 miles east of their chain, and is directly astride the Brothers fault zone. Newberry volcano is essentially an enormous pile of basalt and rhyolite flows that collapsed in the center about at the end of the last ice age. Since then it has been producing rhyolite magmas as obsidian flows and small eruptions of pumice and ash. The most recent eruptions happened within the last few thousand years so it may be reasonable to assume that Newberry volcano is still active. However, large eruptions of obsidian seem to come very late in the careers of volcanoes so it is possible that Newberry may be extinct.

There is evidence all along the west coast of North America of northward movement along faults. The notorious San Andreas fault of California is merely the most conspicuous of many examples. It is hard to be sure when this movement began because the evidence isn't completely clear but there are a lot of indications that it may have been in Miocene time. Certainly that seems to have been the case in Oregon. Evidently the Pacific plate is moving northward, dragging the western edge of the North American continent along.

The descending slab of seafloor beneath the Pacific northwest makes a considerable overlap between the North American and Pacific plates, which must connect them loosely together. This probably explains why the northward movement in southern Oregon is distributed over such a large area. But the earth's crust is probably much thicker in southern than in northern Oregon because the southeastern part of the state seems to be underlain by the buried extension of the Blue Mountains. Therefore, the connection between the continent and the descending slab of seafloor beneath it is probably much tighter in southern than in northern Oregon. If so, the southern part of the state should be moving much more than the northern part and the boundary between those regions could be the Brothers fault zone. North of the Brothers fault zone the northward movement of Oregon appears to be confined mainly to the Coast Range and Willamette Valley. The distance between the two plates is least along the coast and therefore the connection between them is probably tighter there.

Multnomah Falls cascades over the eroded edge of a thick basalt lava flow.

interstate 84

portland — the dalles

This part of the interstate highway follows the narrow gorge the Columbia River has cut through the Cascade Range and the lava plateau beneath it. All of the rocks along the way are dark volcanics, almost all of them are basalt. They make an interesting cross section and reveal some fascinating tidbits of geologic history.

The river has had its problems maintaining a free course through this canyon as various lava flows and giant landslides have tried to block its way. Some of the lava flows, and possibly some of the landslides, have actually dammed the river but it has always managed to erode the obstruction and resume its unfettered course to the ocean. It will be interesting to see what happens in future years now that the river is impounded behind concrete dams and its eroding energy channelled through generating turbines.

The landslides happen because of an interesting arrangement of the bedrock. Deep tropical soils developed in Miocene time mantle an old sequence of plateau basalt flows and are buried beneath a heavy cover of younger basalt flows. Surface water seeps down through vertical cracks in the younger basalt and into the buried soil but can go no deeper because the older basalt below is very nearly watertight. So the buried soil is water saturated and therefore very weak and slippery. The whole sequence is tilted gently southward causing the upper basalt flows on the north side of the river to slide on the slippery wet soil into the Columbia gorge as though they were a stack of heavily buttered pancakes on a tilted platter. That is why all the big landslides are on the Washington side of the river and push its channel towards the Oregon bank. On the Oregon side of the river the buried soil tends to squeeze up through the heavy basalt, covering it like so much toothpaste. In places it is moving the highway, causing what seem to be perpetual repair jobs.

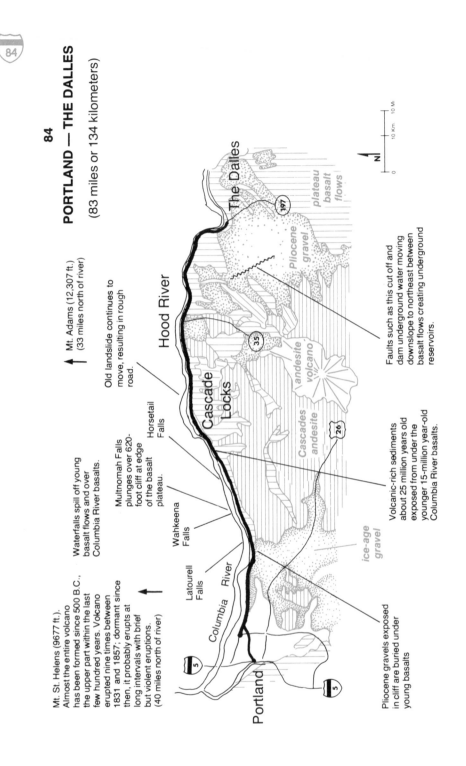

Mt. St. Helens (9677 ft.). Almost the entire volcano has been formed since 500 B.C., the upper part within the last few hundred years. Volcano erupted nine times between 1831 and 1857; dormant since then, it probably erupts at long intervals with brief but violent eruptions. (40 miles north of river)

Waterfalls spill off young basalt flows and over Columbia River basalts.

Multnomah Falls plunges over 620-foot cliff at edge of the basalt plateau.

Mt. Adams (12,307 ft.). (33 miles north of river)

Old landslide continues to move, resulting in rough road.

Faults such as this cut off and dam underground water moving downslope to northeast between basalt flows creating underground reservoirs.

Volcanic-rich sediments about 25 million years old exposed from under the younger 15-million year-old Columbia River basalts.

Pliocene gravels exposed in cliff are buried under young basalts

Latourell Falls

Wahkeena Falls

Horsetail Falls

Columbia River

Portland

Cascade Locks

Hood River

The Dalles

plateau basalt flows

Pliocene gravel

andesite volcano

Cascades andesite

ice-age gravel

The biggest landslide is directly across the river from the Cascade Locks. It involves a section of the Washington bank several miles long extending from North Bonneville eastward to Stevenson. Two smaller slides are visible across the river from the stretch of highway between Cascade Locks and Hood River. The same area also contains still other slides which are no longer moving, so the north bank of the river is almost continuously fringed by active and inactive landslides for a distance of about 20 miles. Numerous patches of highway along the Oregon side of the same section of the river are moving as the buried soil squeezes up beneath them; one of the most persistent of these is about 13 miles west of Hood River.

One of the big lava dams was in the same area as the big Bonneville slide; it extended along several miles of the river east of Cascade Locks. The flow poured into the river gorge from a volcano on the Washington side. A much larger lava dam once existed in the Hood River area; it consisted of several flows erupted from volcanoes on both sides of the river.

Crown Point, about 25 miles east of Portland, is a remnant of a large flood basalt flow that completely filled the canyon of the ancestral Columbia River back in Miocene time, about 20 or 25 million years ago, when the lava plateau was building. Rooster Rock, beside the road directly beneath Crown Point, is a sliver of that ancient flow that slid into the modern canyon of the Columbia River sometime in the very recent geologic past. The scar left where it broke loose is still visible on the cliff above.

East of Crown Point the cliffs of plateau basalt rise higher and higher above the river because the lava plateau is bent upwards into a broad arch beneath the high Cascades. The reason for the arching is a subject for dispute among geologists but many suspect it is due to thermal expansion of the rocks deep beneath the volcanic chain.

In the area about 40 miles east of Portland, just west of the Cascade Locks, the road passes the crest of this arch where the river has cut all the way through the plateau basalt into the older rocks of the western Cascades beneath. Unfortunately, the older rocks are poorly exposed on the Oregon side of the river but there are a few outcrops. These consist of a messy mixture of mudflows and sand obviously washed off an andesite volcano because all the rock fragments are andesite. They and Shellrock Mountain are the only rocks exposed along the road between Portland and the Dalles that are not basalt. Except for

Pressure ridge in a basalt flow. These form when the moving lava wrinkles its solid crust.

the lava dam flows near the Cascade Locks and Hood River, all the basalts at highway level belong to the lava plateau. The younger rocks of the high Cascades are on the mountaintops.

Shellrock Mountain is immediately south of the road just 12 miles west of Hood River. The big piles of shattered rock that cloak its lower flanks crowd right against the highway. They consist of rather coarse-grained rock that looks dark from a distance but shows a dense speckling of white crystals of plagioclase feldspar when examined more closely. It is technically called diorite, and is an igneous rock that intruded the plateau basalts sometime around 10 million years ago and has since been exposed as the Columbia River carved its canyon down through it.

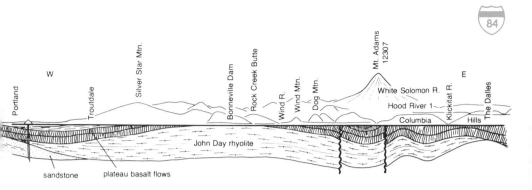

Section along the line of the highway through the Columbia gorge.

A large fault just east of Hood River trends from north to south and has raised the country west of it about 1000 feet during the last few million years. The area east of the Hood River fault is noticeably lower than that to the west and it makes a convenient place to draw a topographic line between the lava plateau and the Cascade Mountains. The narrow gorge of the Columbia River changes at Hood River to a more open canyon that extends to the east, lined on both sides by ledges of plateau basalt.

Grazing cattle trample the soil into rows of miniature terraces. These are on a hillslope about 5 miles south of Biggs Junction.

181

View across the Columbia River towards the basalt cliffs around Biggs.

interstate 84

the dalles — pendleton

Every rock exposed along the route between The Dalles and Pendleton is basalt erupted in floods from the Grand Ronde volcano in northeastern Oregon; geologists nickname it the "Yakima basalt." Between The Dalles and Boardman the highway follows the canyon of the Columbia River past ledges of basalt all stained more or less brown by iron oxides released in weathering. Between Boardman and Pendleton the route crosses the broadly rolling surface of the Columbia lava plateau covered with deep and fertile soil.

The unending cliffs of basalt along the Columbia River clearly show that the same individual lava flows really do continue for miles and miles without varying in any way. No sane geologist would have believed that individual lava flows could possibly cover thousands of square miles and contain hundreds of cubic miles of basalt were it not for the beautiful exposures along the Columbia River. No such eruption has happened during recorded human history.

Their monotonous continuity is the most remarkable thing about these flows. They give no hint of how they were erupted or where they might have come from. For many years geologists surmised that they

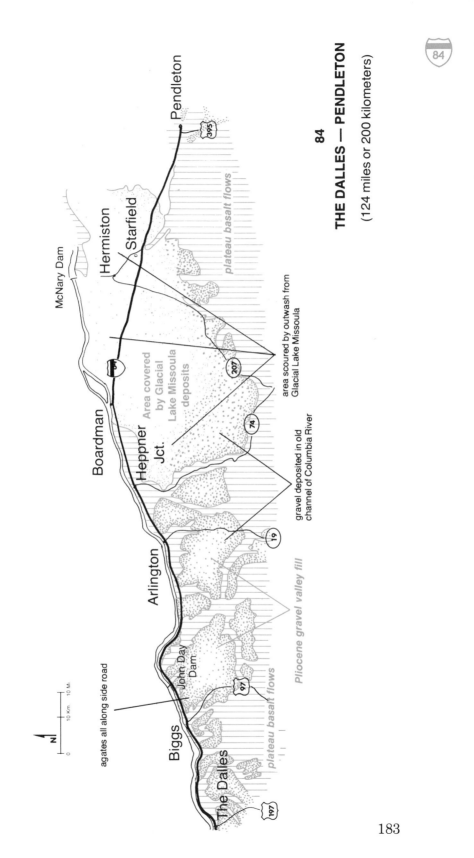

The Dalles

197

Biggs

97

John Day Dam

Arlington

19

Heppner Jct.

74

Boardman

207

McNary Dam

5

Hermiston

Starfield

395

Pendleton

plateau basalt flows

Pliocene gravel valley fill

plateau basalt flows

gravel deposited in old channel of Columbia River

area scoured by outwash from Glacial Lake Missoula

Area covered by Glacial Lake Missoula deposits

agates all along side road

N

0   10 Km. 10 Mi.

84

THE DALLES — PENDLETON

(124 miles or 200 kilometers)

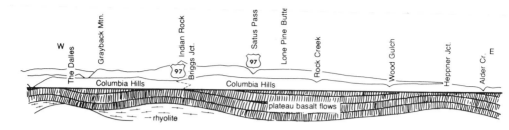

Section along the line of the interstate highway between The Dalles and Pendleton. Basalt lava flows all the way.

might have erupted from long fissures in the ground which were buried under flows from subsequent eruptions and are therefore impossible to find. The Grande Ronde dike swarm in northeastern Oregon has only fairly recently been recognized as their source.

The greatest dam failures of which we have any certain records sent incredible floods down the Columbia River at the end of the last ice age. An enormous glacier dammed the Clark Fork River in northern Idaho backing Glacial Lake Missoula into a large area of western Montana. A dam made of ice is bound to fail eventually and when this one did the whole lake suddenly rushed across eastern Washington and down the Columbia River. This happened at least 35 times as the glacier kept advancing to restore the dam after each catastrophe.

The lake varied in size depending upon how many years the dam managed to last before it burst, so no two floods were quite the same. The largest of them released a wall of water that was 2000 feet high at the dam and several others were almost that big. Of course the floods had lost some of their stature by the time they reached the Columbia River west of Walla Walla but some were still high enough to sweep across the entire area between Boardman and Pendleton. The water surged out of the big bend of the Columbia River and across the plateau surface eroding the ground in some places and covering it with coarse gravel in others. Here and there in this area, unfortunately not near the highway, there are giant ripple marks of gravel big enough to hide a truck. The flooding of this high plateau surface seems incredible at first, but remember that on the days the ice dam burst the Columbia River carried more water for a few hours than all the other rivers of the world combined.

The upper surface of the Columbia lava plateau must have been perfectly level after the last lava flow covered it. It is still remarkably

184

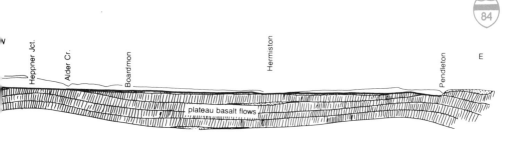

Heppner Jct. Alder Cr. Boardmon Hermiston Pendleton E

plateau basalt flows

level today but there are big swells in the ground that look almost like waves many miles across. And in a way, that is what they are — large, open folds in the lava plateau which reflect the very slight movements that have affected the earth's crust in this area during the last 15 million years.

Roadcut in basalt beside U.S. 97 just south of Biggs Junction.

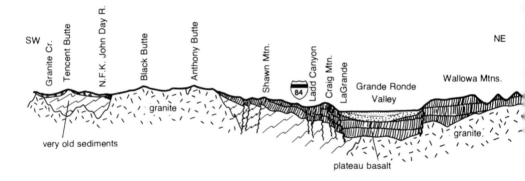

interstate 84

pendleton — baker

The northern part of the route crosses the Columbia lava plateau where all the rocks are basalt and the southern end reaches part of the Blue Mountains which stood above the basalt floods. Rocks in the Blue Mountains are a scrambled mixture of old sediments penetrated by granites and covered in the low places by basalt and very young valley-fill deposits.

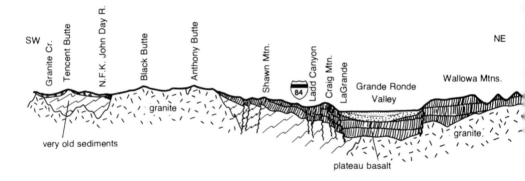

Section across the line of I-80 near LaGrande.

Pendleton is in a valley floored by coarse gravel washed into it during Pliocene time, about 3 to 10 million years ago, when the climate was very dry. The valley itself is almost certainly an old stream valley eroded during Miocene time when the climate was wet and then filled with gravel during Pliocene time.

About 5 miles east of Pendleton the road crosses from the Pliocene gravel onto the basalt plateau which it traverses to a point about midway between LaGrande and Baker. Numerous good roadcuts along the Emigrant Hill grade leading to Deadman Pass expose basalt flows much thinner than those farther west along the Colum-

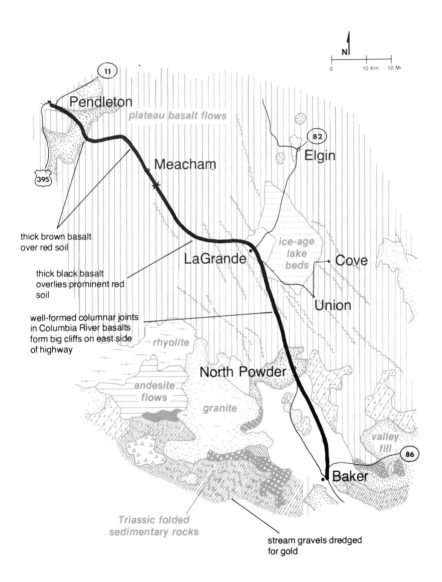

N

0   10 Km.   10 Mi.

11

Pendleton

plateau basalt flows

82

Elgin

Meacham

395

thick brown basalt
over red soil

*ice-age
lake
beds*

LaGrande

Cove

thick black basalt
overlies prominent red
soil

Union

well-formed columnar joints
in Columbia River basalts
form big cliffs on east side
of highway

rhyolite

North Powder

*andesite
flows*

granite

*valley
fill*

86

Baker

*Triassic folded
sedimentary rocks*

stream gravels dredged
for gold

Buried red soil separates two lava flows exposed in a roadcut near Deadman Pass.

bia River. Bright red buried soil horizons separate the flows making them easy to distinguish. It seems likely that they are thin here because this area was on the sloping flank of the old Grande Ronde volcano and thicker farther west because they ponded there on flat ground. The steep slope of Emigrant Hill is due to warping of the lava plateau, not to the original slope on the old Grande Ronde volcano which must have been much flatter. The red soils between flows are laterites which remind us again that Oregon had a wet, tropical climate during Miocene time when flood basalt eruptions built the plateau. The brown stain that colors the Miocene basalts almost certainly formed then because the younger basalts in Oregon are black, as they should be.

View across the lava plateau from Deadman Pass.

View east from La Grande toward the Wallowa Mountains.

The highway crosses from the basalt plateau into the Blue Mountains somewhere near the community of North Powder about halfway between LaGrande and Baker. The exact point of the crossing is hard to spot because the bedrock along that stretch of road is buried beneath more Pliocene valley-fill sediments which also floor the Baker Valley. However, contorted sedimentary rocks intruded by granites make the hills east of North Powder as well as the knobs that rise above the floor of the Baker Valley west of the highway. A building and monument-stone quarry for years produced gray granite attractively freckled with dark spots from one of those knobs.

Pressure ridge in basalt beside the road about 15 miles east of Bend.

u.s. 20

bend — burns

Except for the obsidian at Glass Buttes, all the rocks exposed along the road between Bend and Burns are basalt lava flows erupted within the last few million years. Many of them are quite young indeed and still fairly fresh and unweathered. This is bleak and nearly treeless country with very thin soils so the rocks are easy to see.

Between Bend and Millican the road passes north of Newberry volcano. Although it is more visible from U.S. 20 than from any other road, Newberry volcano still doesn't make much of an outline against

190

BEND — BURNS

(130 miles or 207 kilometers)

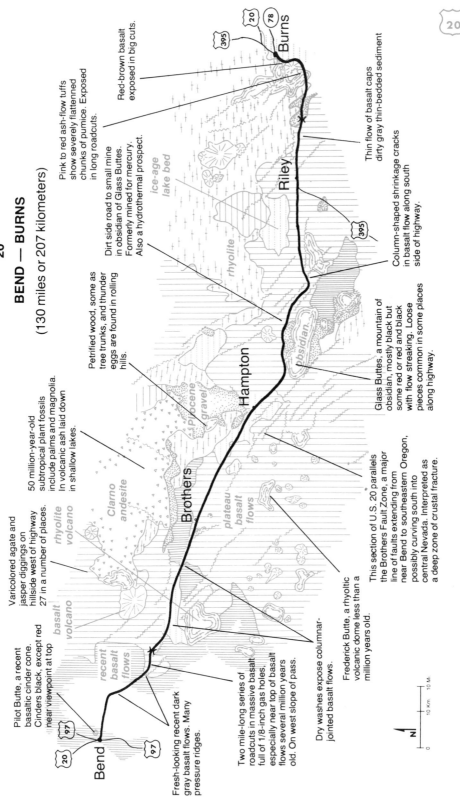

Pink to red ash-flow tuffs show severely flattenned chunks of pumice. Exposed in long roadcuts.

Red-brown basalt exposed in big cuts.

Thin flow of basalt caps dirty gray thin-bedded sediment

Dirt side road to small mine in obsidian of Glass Buttes. Formerly mined for mercury. Also a hydrothermal prospect.

Column-shaped shrinkage cracks in basalt flow along south side of highway.

Petrified wood, some as tree trunks, and thunder eggs are found in rolling hills.

50 million-year-old subtropical plant fossils include palms and magnolia. In volcanic ash laid down in shallow lakes.

Glass Buttes, a mountain of obsidian, mostly black but some red or red and black with flow streaking. Loose pieces common in some places along highway.

Varicolored agate and jasper diggings on hillside west of highway 27 in a number of places.

This section of U.S. 20 parallels the Brothers Fault Zone, a major line of faults extending from near Bend to southeastern Oregon, possibly curving south into central Nevada. Interpreted as a deep zone of crustal fracture.

Pilot Butte, a recent basaltic cinder cone. Cinders black, except red near viewpoint at top

Frederick Butte, a rhyoltic volcanic dome less than a million years old.

Dry washes expose columnar-jointed basalt flows.

Two mile-long series of roadcuts in massive basalt full of 1/8-inch gas holes, especially near top of basalt flows several million years old. On west slope of pass.

Fresh-looking recent dark gray basalt flows. Many pressure ridges.

ice-age lake bed

rhyolite

obsidian

Pliocene gravel

plateau basalt flows

Clarno andesite

rhyolite volcano

basalt volcano

recent basalt flows

Riley

Burns

Hampton

Brothers

Bend

0   10 Km   10 Mi.

N

the sky despite its great size. Nevertheless, it is an immense pile of rock, mostly basalt lava flows which run out thin and flat instead of piling up to make a steep mountain as andesite does.

The flanks of Newberry volcano are broken out in a rash of basalt cinder cones like a giant case of warts. They dot the landscape south of the highway with the biggest being China Hat which is east of Newberry volcano and almost straight south of Millican. Many of them are very young. This kind of thing is typical of large volcanoes in their later stages of activity when they tend to erupt through a scattering of cinder cones on their flanks instead of through the main central vent. Each cinder cone marks a single eruption which normally begins with construction of a cone of bubbly basalt fragments coughed out of the vent and ends with the appearance of one or two lava flows which burst through the base of the loose pile of cinders. Once a cinder cone quits, and erupts its lava flow, it is out of business; they are one-shot volcanoes which never erupt a second time.

Between Millican and Riley, U.S. 20 follows the Brothers fault zone which is named for the little community of Brothers. The zone is a swarm of faults that trend approximately parallel to the road and slip sideways with the south side moving northwest. The fact that the faults move horizontally instead of vertically minimizes their effect on the landscape. But they do have an effect just the same, creating some roughness through slippage of long slivers of the earth's crust along faults.

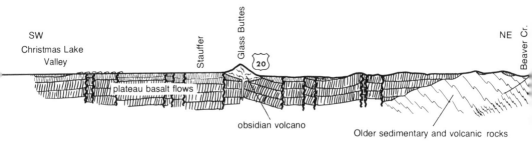

Section across the line of U.S. 20 at Glass Buttes.

Glass Buttes is a rare and extraordinary sight, a whole mountain of obsidian. It is essentially an oversized lava flow which was so thick and viscous that it piled up around the vent to make a mountain instead of spreading thinly across the countryside. The Glass Buttes mass can't be very old but it has been there long enough to be sliced up a bit by movement on several faults.

Chunks of streaky brown obsidian in a talus slope at Glass Buttes.

Obsidian is a variety of rhyolite which forms when the magma cools to become a glass without crystallizing. Its extremely low water content is the only thing besides its glassiness that distinguishes obsidian from ordinary rhyolite. Evidently it must form from rhyolite magmas that somehow manage to melt and make it all the way to the surface without absorbing any water. But there is no obvious explanation for the fact that only a few rhyolite magmas stay dry and become obsidian while most of them absorb water before they reach the surface and erupt as pumice or ash.

Much of the Glass Buttes obsidian is beautifully streaked with red or brown and is much handsomer than ordinary obsidian which is just plain black. No rock collector can resist hauling a chunk along. Fortunately, the mountain is big enough that it seems unlikely to be endangered for a long time to come.

u.s. 20

burns — ontario

Most of the route crosses an especially bleak portion of the lava plateau, a high and windswept desert. All of the hard rock along the way is basalt. It is thinly covered by soil on the high plateau and deeply buried in the Ontario area by thick sedimentary deposits that make excellent soils.

Burns is at the northwest corner of the Harney Valley which is almost square in outline except for Harney Lake which fills a much smaller circular annex at its southwest corner. It seems likely that this smaller, circular valley may be a caldera, a basin formed when the surface subsided as molten material was withdrawn from beneath during volcanic eruptions. It is about the usual size and shape of such features and is in a region that has seen a great deal of volcanic activity. Some geologists suspect that Burns may be in another caldera whose outlines are now so blurred by later geologic events that they are no longer visible in the landscape.

For 21 miles between Burns and Buchanan the road follows an absolutely straight course across the valley surface levelled to billiard table standards of flatness. No other geologic process except

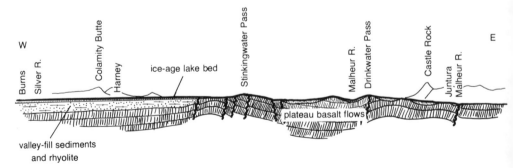

Section along the line of U.S. 26 from Burns to Ontario.

deposition in the bed of a lake is likely to create such a perfectly flat surface.

On the east side of the Harney Valley the road crosses a steep fault scarp onto the high surface of the Harney plateau which it crosses between Buchanan and Juntura. A rhyolite flow exposed in the base of that scarp near Buchanan contains some excellent thundereggs.

The top of the Harney plateau is harsh country with ragged scabs of black basalt poking through the thin deposits of pale gray silt that the wind has blown onto parts of the landscape. Movements along several faults that trend approximately at right angles to the road have shoved blocks of this plateau up and down to define the basic framework of the landscape. Erosion hasn't made much of a mark on this scene of lava flows offset by faults, partly because the faulting is very recent and partly because the flows are so porous they absorb surface water and inhibit development of streams.

Between Juntura and Ontario, U.S. 20 follows the valley of the Malheur River which flows eastward off the Harney plateau to the Snake River. Along most of the way between Juntura and Vale the river has cut a deep and winding canyon into the plateau, making beautiful cross-sections of lava flows in the valley walls. In a number of places these are rather flamboyantly marked by old soils buried between flows which show up as horizontal red streaks in the canyon wall. Their color is especially interesting because it is typical of soils that form in wet, tropical climates such as prevailed in Oregon back in Miocene time when these flows erupted.

Between the area a few miles west of Vale and Ontario the road crosses a broad valley deeply floored by deposits of sediment. Some of this accumulated in a large lake impounded behind the big lava flows

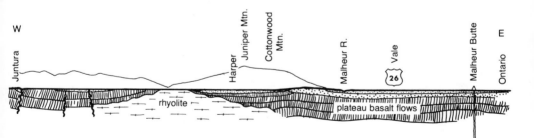

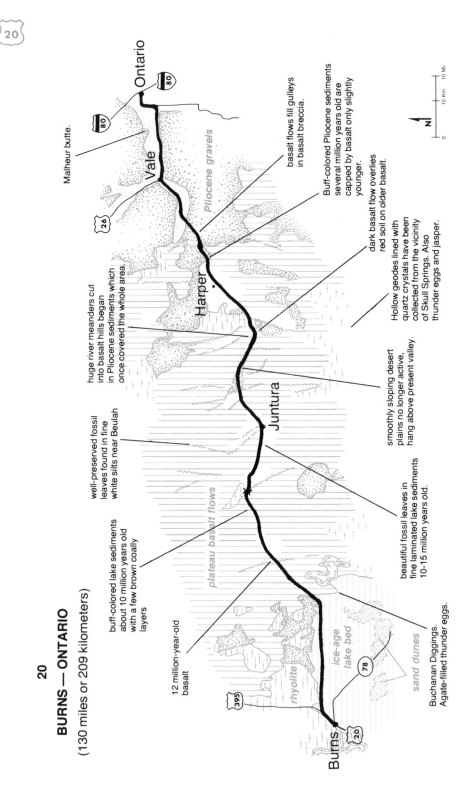

BURNS — ONTARIO

(130 miles or 209 kilometers)

Malheur butte.

Pliocene gravels

Vale

basalt flows fill gulleys in basalt breccia.

Buff-colored Pliocene sediments several million years old are capped by basalt only slightly younger.

dark basalt flow overlies red soil on older basalt.

Harper

Hollow geodes lined with quartz crystals have been collected from the vicinity of Skull Springs. Also thunder eggs and jasper.

huge river meanders cut into basalt hills began in Pliocene sediments which once covered the whole area.

smoothly sloping desert plains no longer active, hang above present valley.

Juntura

well-preserved fossil leaves found in fine white silts near Beulah

plateau basalt flows

buff-colored lake sediments about 10 million years old with a few brown coally layers

beautiful fossil leaves in fine laminated lake sediments 10-15 million years old.

12 million-year-old basalt

ice-age lake bed

rhyolite

sand dunes

Buchanan Diggings. Agate-filled thunder eggs.

Burns

N

0   10 Km.   10 Mi.

that erupted during Miocene time. These lake beds, which don't outcrop near the road, contain numerous fossil leaves of kinds that grow only in wet, tropical climates, such things as avocado trees. They must have grown on those red soils now buried between lava flows. In most places the lake beds are buried beneath much younger sediments washed into the valley during Pliocene time, between 3 and 10 million years ago, when the climate was very dry and there were no streams to carry the debris of erosion away.

Malheur Butte, a small basalt volcano in the river valley about 5 miles west of Ontario.

u.s. 26

madras — john day

The highway between Madras and John Day crosses some of the most interesting rocks in Oregon. Many of them are older than the flood basalts that cover most of the country to the north and they give us our best clue to what may be hidden beneath the basalt plateau. The area is a window looking back into the more distant past of central Oregon.

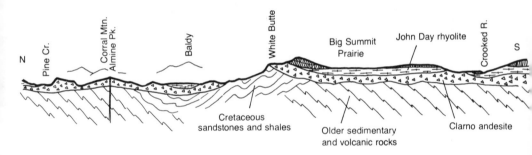

Section across the line of U.S. 26 in the Ochoco Mountains showing the Clarno Formation overlain by the John Day rhyolite ash and the Picture Gorge basalt lava flows.

Madras is on fairly inconspicuous deposits of sedimentary rock consisting of volcanic debris reworked by streams within the last few million years. The massive ledges just west of town are much younger basalt lava flows.

The route from Madras through Prineville to the Ochoco reservoir passes numerous outcrops of pale yellowish-brown rock consisting of rhyolitic volcanic ash. These are the John Day Formation, famous for the numerous wonderful collections of fossil plants and vertebrate animals it has produced during the last century. The formation consists of vast deposits of rhyolite ash erupted from the southern

MADRAS—JOHN DAY
(146 miles or 234 kilometers)

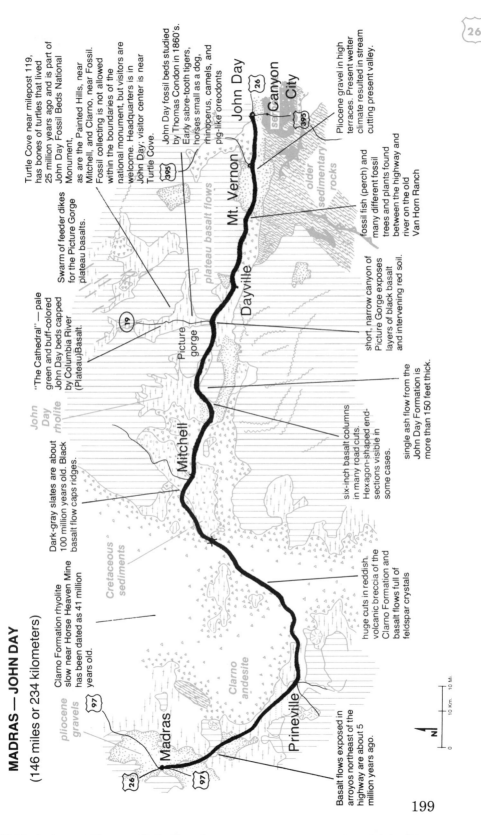

Turtle Cove near milepost 119, has bones of turtles that lived 25 million years ago and is part of John Day Fossil Beds National Monument, as are the Painted Hills, near Mitchell, and Clarno, near Fossil. Fossil collecting is not allowed within the boundaries of the national monument, but visitors are welcome. Headquarters is in John Day; visitor center is near Turtle Cove.

John Day fossil beds studied by Thomas Condon in 1860's. Early sabre-tooth tigers, horses small as a dog, rhinocerus, camels, and pig-like oreodonts

Pliocene gravel in high terraces. Present wetter climate resulted in stream cutting present valley.

Swarm of feeder dikes for the Picture Gorge plateau basalts.

plateau basalt flows

older sedimentary rocks

fossil fish (perch) and many different fossil trees and plants found between the highway and river on the old Van Horn Ranch

"The Cathedral" — pale green and buff-colored John Day beds capped by Columbia River (Plateau)Basalt.

John Day rhyolite

Picture gorge

short, narrow canyon of Picture Gorge exposes layers of black basalt and intervening red soil.

Clarno Formation rhyolite flow near Horse Heaven Mine has been dated as 41 million years old.

Dark-gray slates are about 100 million years old. Black basalt flow caps ridges.

six-inch basalt columns in many road cuts. Hexagon-shaped end-sections visible in some cases.

Cretaceous sediments

single ash flow from the John Day Formation is more than 150 feet thick.

huge cuts in reddish, volcanic breccia of the Clarno Formation and basalt flows full of feldspar crystals

Clarno andesite

pliocene gravels

Basalt flows exposed in arroyos northeast of the highway are about 5 million years ago.

Mitchell Rock punctuates a hillside north of the road about a mile west of town. It is an old volcanic neck, part of the Clarno formation.

part of the western Cascades during late Oligocene and early Miocene time, perhaps 20 to 30 million years ago. Most of the good thunderegg collecting localities are in the John Day Formation, including several in the Prineville area.

An early citizen of Oregon who was buried in volcanic ash of the John Day formation about 25 million years ago.

Rocks exposed in the 30-mile stretch through the Ochoco Mountains between the Ochoco reservoir and the area about 5 miles west of Mitchell are dark andesites and basalts erupted during late Eocene and early Oligocene time, between about 45 and 35 million years ago. They belong to the Clarno Formation. The Ochoco Mountains are really an eastern extension of the older western Cascades, a volcanic chain that followed the trend of the Cascades as far north as Eugene and then curved east through central Oregon at least as far as the area just south of Pendleton. Unlike the western Cascades, the Ochocos are full of clearly recognizeable volcanos, presumably because no later activity has buried them.

The highway crosses some of the most interesting rocks in central Oregon for about 5 miles around Mitchell. They are so ugly that only a geologist could love them; dull gray and brown shales and pebbly sandstones that make inconspicuous outcrops. The shales are full of round concretions of harder rock which often contain fossils, usually the remains of extinct relatives of the pearly nautilus called ammonites, which lived in coiled shells that look almost as though they might be oversized snail shells. The fossils tell us these rocks were deposited in seawater during late Cretaceous time, about 75 million years ago.

At least 5000 feet of these marine sediments are known to exist in the Mitchell area and their base is not exposed, so the actual thickness may be much greater. It seems very likely that much of north-central Oregon may be underlain by great thicknesses of such rocks. These are precisely the kinds of rocks that produce oil and gas, so the outcrops along the road west of Mitchell offer hope of future petroleum production from somewhere beneath the lava plateau in central Oregon. The areas of Cretaceous sedimentary rock are easy to spot because they erode into broad valleys with smooth floors that make good farmland. The younger volcanic rocks in the neighborhood make much rougher country covered with sagebrush or juniper and fit mostly for grazing.

The 32 miles of road between Mitchell and Picture Gorge passes more exposures of dark, andesitic rubble belonging to the Clarno Formation and a number of plateau basalt lava flows exposed as prominent dark ledges in the valley walls above the river. No fewer than 17 of these flows are cleanly cross-sectioned in the steep walls of Picture Gorge where they have been very carefully studied. They are all the distinctive kind of basalt generally found south of the Blue Mountains, the so-called "Picture Gorge basalt." These flows erupted from dikes in the area between Monument and Dayville just east of Picture Gorge. They are part of an older and probably smaller volcano than the one that produced the Columbia plateau basalts north of the Blue Mountains. Careful study of the flows in Picture Gorge was one of the important steps that led geologists to realize they could distinguish basalt flows from different sources and subdivide the lava plateau into parts.

Picture Gorge is a spectacular canyon cut through a thick sequence of basalt lava flows. It is always marvelous to see a stream flow for miles through fairly soft rock and then suddenly pass through a narrow gorge laboriously carved in much harder rock; especially when a short detour would have kept it on softer rocks. It is not true that streams always choose a path of least resistance. In this case the explanation is easy to see.

Many hilltops in the John Day country are capped by loose sands and gravels laid down during Pliocene time, about 5 million years ago give or take a few million. These are the youngest deposits in the area and they are on top of everything else, a brick-red ledge of welded ash right in the middle of the section makes them easy to spot. Those sands and gravels must have been deposited on the land surface that existed during Pliocene time at the level of the present hilltops before

Cathedral Rock is eroded in the pale volcanic ash of the John Day formation. It is near Oregon 19 a few miles north of Picture Gorge.

Hills in the pale volcanic ash north of Dayville. Most of this material is rhyolite erupted from the Strawberry Mountains.

the modern stream valleys existed. So we can imagine the modern streams beginning to flow along their present courses on the freshly deposited Pliocene sediments at the level of the modern hilltops. They probably got started about 3 million years ago when the first ice age began and the climate of the entire northwest seems to have become much wetter than it had been during Pliocene time. Erosion has since brought the rivers to their present levels and the landscape to its modern appearance. So the John Day River began flowing on soft Pliocene sediments, not realizing that it would shortly erode its way down onto a buried hill made of hard basalts and be forced to cut Picture Gorge.

The long range south of the highway between Dayville and John Day is the Aldrich Mountains, still another window into the pre-flood-basalt past of central Oregon. Most of their rocks are a mess of old seafloor sediments deposited between 250 and 150 million years ago and then scraped off onto the continent as the seafloor sank into the earth's interior beneath them. Parts of the Aldrich Mountains, especially the area south of John Day, also contain large masses of serpentinite and black peridotite which must be slices of the oceanic crust that somehow got mixed into the mountains instead of sinking back into the earth.

So the Aldrich Mountains are a typical coastal mountain range just like the Klamaths except that they aren't near the coast. They mark the line of seafloor sinking that existed about 250 to 150 million years ago. The fact that some of their sedimentary rocks are full of small andesite fragments means that a chain of volcanoes must have existed then too. Presumably these are in the area south of Aldrich Mountains, now covered by an opaque crust of flood-basalt flows.

Picture Gorge, the John Day River cut this gash through a section of hard basalt flows.

u.s. 26

john day — vale

Except for a very small patch of old seafloor sediments near Dixie Pass and another about 5 miles east of Ironside, all the solid bedrock exposed along the road between John Day and Vale is volcanic rock erupted at different times from at least 3 sources.

The highway follows the broad valley of the John Day River for 13 miles east of town before it starts up the slope to Dixie Pass. Rocks north of this stretch of road are basalt lava flows erupted sometime during Miocene time, about 15 or 20 million years ago. They most likely came from the Picture Gorge volcano. The long slopes that slant smoothly upwards to the mountains south of the road are underlain by gravels that washed out of the mountains and formed an apron of coalescing alluvial fans at their base. That sort of thing happens in deserts so those slopes must be souvenirs of a period when the climate was much drier than it is now, probably Pliocene time between about 3 and 10 million years ago. Some placer gold has come out of those gravels.

Between John Day and Dixie Pass, the road crosses messy gray andesites erupted in late Eocene and early Oligocene time, about 45 to 35 million years ago, when an early version of the Cascade Range curved east from the Eugene area and extended through central Oregon. The coast and the line of seafloor sinking also curved eastward then so these volcanic rocks erupted just as near the seashore as the modern Cascades. They belong to the Clarno Formation which also makes much of the Ochoco Mountains between John Day and Prineville.

The Ochoco andesites covered older rocks which erosion had already carved into a hilly landscape. These are old seafloor sediments and chunks of the bedrock seafloor that had been scrambled into the

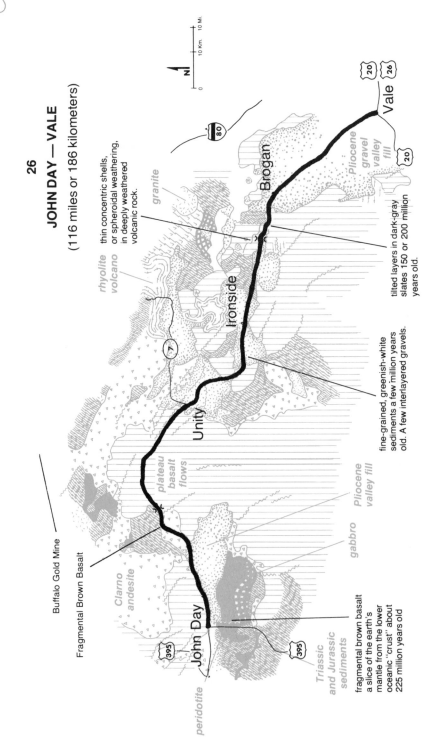

thin concentric shells, or spheroidal weathering, in deeply weathered volcanic rock.

granite

rhyolite volcano

tilted layers in dark-gray slates 150 or 200 million years old.

fine-grained, greenish-white sediments a few million years old. A few interlayered gravels.

Buffalo Gold Mine

Fragmental Brown Basalt

plateau basalt flows

Pliocene valley fill

gabbro

Clarno andesite

Triassic and Jurassic sediments

peridotite

fragmental brown basalt a slice of the earth's mantle from the lower oceanic "crust" about 225 million years old

Pliocene gravel valley fill

Vale

Brogan

Ironside

Unity

John Day

Blue Mountains along the line of seafloor sinking. The top of one of those buried hills, now exposed again by erosion, is exposed in a large area north of the road and a small area south of the road between the John Day River and Dixie Pass. The road crosses it in a very small area with few exposures just east of Dixie Pass. The rocks are basically similar to those that make Canyon Mountain southeast of John Day; those exposed along the road are very dark mudstones with a distinctly slaty look.

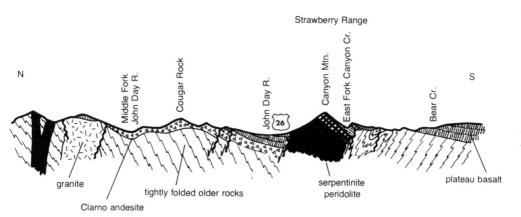

Section across U.S. 26 just east of John Day.

Between Dixie Pass and Unity Reservoir, U.S. 26 crosses a broad tract of andesites erupted from a large volcanic center in the Strawberry Mountains southeast of John Day. Although the Strawberry volcanics contain some basalt, they seem to consist mostly of andesite and an enormous quantity of light-colored ash spread over quite a large region. The Strawberry volcanics erupted in an area between the Picture Gorge and Grande Ronde volcanoes during the period after the activity of the Picture Gorge volcano had ceased and before that of the Grande Ronde volcano began. The appearance of large volumes of andesite in this place at that time is most unusual and difficult to understand. Andesite normally erupts from volcanic chains parallel to a line of seafloor sinking and rarely from an isolated center well inland from the coast. Remember that by Miocene time the line of seafloor sinking was already about where it is today.

Between Unity Reservoir and Brogan, U.S. 26 crosses a series of low hills mostly underlain by more Strawberry volcanics and broad valleys floored with gravels and volcanic ash washed into them during times of drier climate when the stream drainage wasn't pow-

These elegantly geometric shrinkage columns are in an andesite lava flow exposed near Dixie Pass. It is part of the strawberry volcanics.

erful enough to carry sediment away, probably during Pliocene time. Where the valley fill is eroding in the wetter modern climate, it makes low, softly-rounded hills sparsely covered with vegetation. Sagebrush thrives on the stuff.

A small group of low hills about 5 miles east of Ironside contains more old seafloor sediments and volcanics basically similar to those near Dixie Pass. This is another old hilltop of the Blue Mountains long buried beneath younger rocks and now exposed again by more recent erosion. There aren't many good outcrops, though, and not much to see.

The highway follows the floodplain of Willow Creek between Brogan and Vale. The low hills on either side of the floodplain are eroded in valley-fill sediments that accumulated several million years ago when the climate was much drier and are eroding now that the climate is wetter.

Bluffs in white rhyolitic volcanic ash about 17 miles north of Jordan Valley.

u.s. 95

ontario — mcdermitt

This is desolate and unpeopled country with broad valleys and craggy hills spread beneath an endless sky. Every outcrop of bedrock along the way is volcanic, either dark basalt or pale rhyolite. Most of the basalt is lava flows and most of the rhyolite volcanic ash but some of the ash beds are solidly enough welded to make strong ledges.

Valley-fill deposits floor the low parts of the landscape. Many of these are actively accumulating today, because this is arid country in which the streams don't carry enough water to haul away all the debris of erosion. Indeed, many valleys are quite undrained and contain stale lakes of alkaline and salty water.

The fact that the plateau volcanic rocks in southeastern Oregon include both rhyolite and basalt, instead of just basalt as is the case farther north, probably tells us something about differences in the composition of the underlying crust. Basalt forms when the black rocks of the earth's interior partially melt, usually through relief of pressure. The only thing that keeps things solid at the high temperatures that prevail down there is the pressure of the rocks above. Similarly, rhyolite can form by pressure-relief melting of the light-colored rocks that make up the continental crust.

JORDAN VALLEY — MCDERMITT

(101 miles or 164 kilometers)

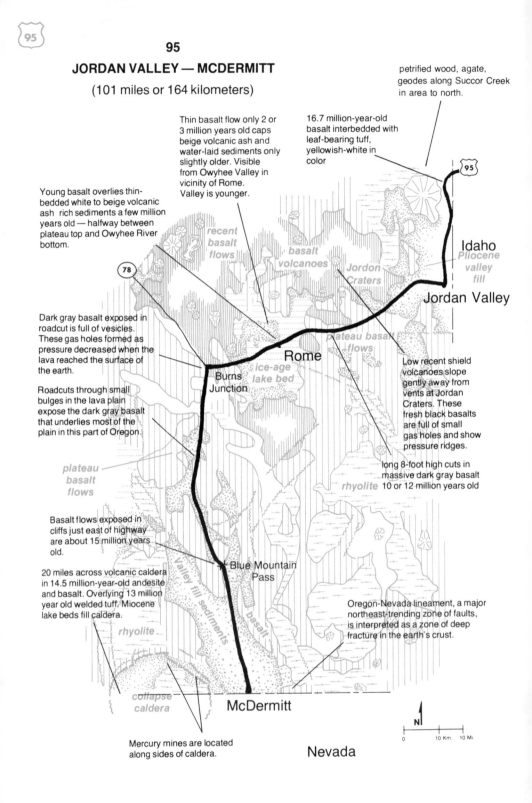

petrified wood, agate, geodes along Succor Creek in area to north.

Thin basalt flow only 2 or 3 million years old caps beige volcanic ash and water-laid sediments only slightly older. Visible from Owyhee Valley in vicinity of Rome. Valley is younger.

16.7 million-year-old basalt interbedded with leaf-bearing tuff, yellowish-white in color

Young basalt overlies thin-bedded white to beige volcanic ash rich sediments a few million years old — halfway between plateau top and Owyhee River bottom.

Idaho

Pliocene valley fill

recent basalt flows

basalt volcanoes

Jordon Craters

78

Jordan Valley

Dark gray basalt exposed in roadcut is full of vesicles. These gas holes formed as pressure decreased when the lava reached the surface of the earth.

plateau basalt flows

Rome

ice-age lake bed

Burns Junction

Low recent shield volcanoes slope gently away from vents at Jordan Craters. These fresh black basalts are full of small gas holes and show pressure ridges.

Roadcuts through small bulges in the lava plain expose the dark gray basalt that underlies most of the plain in this part of Oregon.

long 8-foot high cuts in massive dark gray basalt 10 or 12 million years old

plateau basalt flows

rhyolite

Basalt flows exposed in cliffs just east of highway are about 15 million years old.

Valley fill sediments

Blue Mountain Pass

20 miles across volcanic caldera in 14.5 million-year-old andesite and basalt. Overlying 13 million year old welded tuff. Miocene lake beds fill caldera.

basalt

Oregon-Nevada lineament, a major northeast-trending zone of faults, is interpreted as a zone of deep fracture in the earth's crust.

rhyolite

collapse caldera

McDermitt

N

0   10 Km.   10 Mi.

Mercury mines are located along sides of caldera.

Nevada

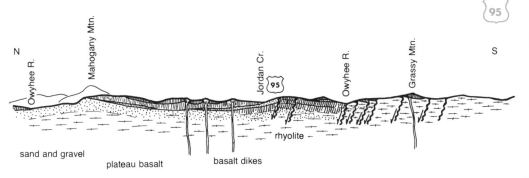

Section across the line of the highway near Arock.

There is no continental crust beneath northern Oregon; the sediments and volcanics there rest on the slab of oceanic crust that quit moving when the line of seafloor sinking shifted to its present position about 35 million years ago. So there is nothing deep below the surface in northern Oregon that could turn into rhyolite. There is continental crust beneath southeastern Oregon, probably rocks similar to those in the Blue and Klamath Mountains. Such rocks are actually exposed in the Pueblo Mountains about 50 miles west of McDermitt. So when the events of Miocene time stretched the crust and relieved pressure on the hot rocks at depth, only basalt came to the surface in northern Oregon and both basalt and rhyolite in southeastern Oregon.

These sediments exposed along the east side of the Owyhee River near Rome consist mostly of light-colored volcanic ash washed into the valley floor.

211

The rimrock is thin flows of basalt on top of thick deposits of light-colored volcanic ash. Beside U.S. 95 about 5 miles east of Burns Junction.

Highway 95 crosses the Brothers fault zone in the area between Jordan Valley and Burns Junction. Its main attraction in this part of Oregon is a spectacular field of fresh basalt cinder cones and lava flows in the area north of the highway and accessible from Arock. The rocks look fresh but they do have some sagebrush on them so it is easy to believe that they are actually several thousand years old.

The route between Burns Junction and McDermitt continues to cross basalt lava flows and light-colored rhyolite ash. Most of this is plateau volcanics erupted during Miocene time although there are some younger basalts in the area away from the road southeast of Burns Junction.

All of southeastern Oregon has a very strange landscape created almost entirely by faulting and hardly at all by erosion. The volcanic activity must have left a fairly flat and characterless scene covered by basalt flows. Crustal movements have since broken this into an uncounted number of fault blocks of all sizes. Some of them are mountain ranges, others amount to no more than big steps on a hillside. Nearly all the high and low places in the landscape are fault blocks moved up or down in the very recent past; most of them are probably still moving. The processes of erosion have merely washed some sediment into the low places and carved occasional stream valleys.

212

Coarse gravels deposited on top of the lava plateau during Pliocene time when the climate was very dry.

u.s. 97

biggs junction — bend

The northern half of the route between Biggs Junction and Bend crosses the gently undulating surface of the Columbia Plateau, a sea of basalt. The southern half crosses a variety of volcanic rocks, some older than the plateau basalts and a few younger.

Biggs is in the deep canyon the Columbia River has eroded for itself during the last few million years. South of town the road winds laboriously up the valley wall until it finally tops out onto the broad surface of the lava plateau which stretches away into the distance as though it would never end.

For several miles both north and south of Wasco, the road crosses gravels deposited on the lava plateau during Pliocene time, roughly between 3 and 10 million years ago. The gravel deposits on this part of the Columbia plateau form a long line about 10 miles wide that runs approximately parallel to the Columbia River from the area

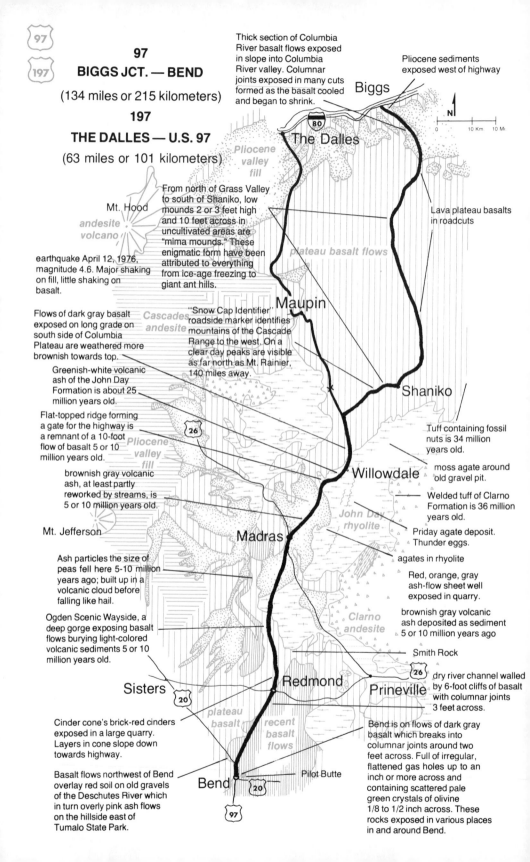

97

BIGGS JCT. — BEND

(134 miles or 215 kilometers)

197

THE DALLES — U.S. 97

(63 miles or 101 kilometers)

Thick section of Columbia River basalt flows exposed in slope into Columbia River valley. Columnar joints exposed in many cuts formed as the basalt cooled and began to shrink.

Pliocene sediments exposed west of highway

Biggs

The Dalles

Pliocene valley fill

plateau basalt flows

Lava plateau basalts in roadcuts

Mt. Hood

andesite volcano

earthquake April 12, 1976, magnitude 4.6. Major shaking on fill, little shaking on basalt.

From north of Grass Valley to south of Shaniko, low mounds 2 or 3 feet high and 10 feet across in uncultivated areas are "mima mounds." These enigmatic form have been attributed to everything from ice-age freezing to giant ant hills.

Maupin

Flows of dark gray basalt exposed on long grade on south side of Columbia Plateau are weathered more brownish towards top.

Cascades andesite

"Snow Cap Identifier" roadside marker identifies mountains of the Cascade Range to the west. On a clear day peaks are visible as far north as Mt. Rainier, 140 miles away.

Greenish-white volcanic ash of the John Day Formation is about 25 million years old.

Flat-topped ridge forming a gate for the highway is a remnant of a 10-foot flow of basalt 5 or 10 million years old.

Pliocene valley fill

brownish gray volcanic ash, at least partly reworked by streams, is 5 or 10 million years old.

Shaniko

Tuff containing fossil nuts is 34 million years old.

moss agate around old gravel pit.

Willowdale

Welded tuff of Clarno Formation is 36 million years old.

John Day rhyolite

Priday agate deposit. Thunder eggs.

Mt. Jefferson

Madras

agates in rhyolite

Ash particles the size of peas fell here 5-10 million years ago; built up in a volcanic cloud before falling like hail.

Red, orange, gray ash-flow sheet well exposed in quarry.

brownish gray volcanic ash deposited as sediment 5 or 10 million years ago

Clarno andesite

Ogden Scenic Wayside, a deep gorge exposing basalt flows burying light-colored volcanic sediments 5 or 10 million years old.

Smith Rock

Sisters

Redmond

Prineville

dry river channel walled by 6-foot cliffs of basalt with columnar joints 3 feet across.

plateau basalt

recent basalt flows

Cinder cone's brick-red cinders exposed in a large quarry. Layers in cone slope down towards highway.

Bend is on flows of dark gray basalt which breaks into columnar joints around two feet across. Full of irregular, flattened gas holes up to an inch or more across and containing scattered pale green crystals of olivine 1/8 to 1/2 inch across. These rocks exposed in various places in and around Bend.

Basalt flows northwest of Bend overlay red soil on old gravels of the Deschutes River which in turn overly pink ash flows on the hillside east of Tumalo State Park.

Bend

Pilot Butte

south of Hermiston west to the Cascades. We believe this is an old gravel-filled valley that the Columbia flowed through before Pliocene time. There are no Pliocene gravel deposits in the canyon the Columbia now follows which probably means that the river has carved that canyon during the 3 million years since the end of Pliocene time.

There aren't many rocks to look at along the route across the top of the Columbia plateau between Biggs and Willowdale. The entire surface is underlain by flood basalt lava flows erupted about 15 or 20 million years ago.

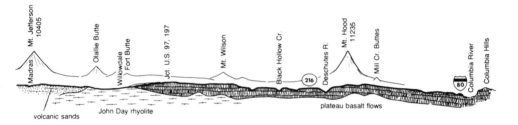

Section from Madras to the Columbia River.

The basalt flows are very solid rocks almost lacking pore space, but they do have rubbly zones and old soil horizons above and below them which are porous and contain a lot of water. Water wells drilled deep enough in the lava plateau are almost certain to cut these water-bearing zones which are often lavishly productive. Deep-well irrigation is increasing rapidly because the dryland soils of the Columbia Plateau respond prolifically to a bit of water.

The danger is that deep-well irrigation, like any other good thing, is too easily overdone. The water-bearing zones in the Columbia Plateau basalts don't refill very rapidly so it doesn't take very many irrigation projects to pump water faster than rainfall can replace it. When that perilous line is crossed, the reservoir of stored ground water becomes an unrenewable resource, for all practical purposes, and irrigation can continue only as long as the water lasts. This puts farming on the same kind of boom and bust basis as mining or any other industry based on a depleting resource. The kind of bust that permanent exhaustion of the irrigation water would cause is worse than the kind farmers are now used to.

About 3 miles north of Willowdale U.S. 97 leaves the flood basalts and crosses onto the older volcanic rocks of the Ochoco Mountains. In the Willowdale area these consist mostly of the John Day Formation, a thick accumulation of light-colored volcanic ash deposits erupted from Cascade volcanoes about 25 to 30 million years ago. There are some older volcanic rocks, the Clarno Formation a few miles back in the hills, and about 15 miles south of Willowdale there is a patch of crumpled sedimentary rocks similar to those in the Blue Mountains. Evidently such rocks must lie buried beneath a large part of central Oregon.

Most of the volcanic ash in the John Day Formation was cold by the time it settled to the ground to make loose deposits that might drift in the wind or wash in the rain before they finally solidified enough to be stable. These deposits are still fairly soft rock which weathers to softly rounded hills in which the bedrock reveals itself mainly in patches of pastel color and occasional outcrops in gullies or creek beds. But some of the ash arrived still partially melted, and welded itself into solid rock the instant it touched the ground. These sheets of welded ash make the occasional strong ledges and cliffs that add touches of ruggedness to the hills on the John Day Formation. Central Oregon must have been fairly lush country when these ash deposits accumulated because we find them full of fossil wood, leaves, and bone that have delighted scientists and collectors alike for generations.

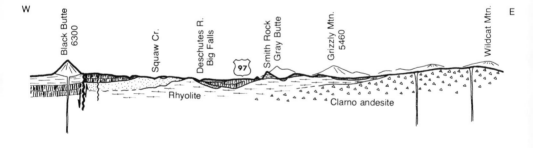

Section across U.S. 97 just north of Redmond.

The county road to Ashwood that takes off from U.S. 97 a mile or so south of Willowdale passes the Priday Ranch which is famous among agate fanciers everywhere as the source of thousands of beautiful thundereggs. They come from a ledge of partially welded rhyolite ash in the John Day Formation.

Thundereggs, Oregon's official state rock, are unusual and distinctive agates found only in light-colored volcanic ash at a number of localities in Oregon and elsewhere. Like all agates, thundereggs form as circulating ground water slowly fills cavities in the original rock with silica. So agates are not volcanic even though they are usually quite common in volcanic rocks, simply because those tend to have lots of open cavities and are good sources of dissolved silica.

Although each thunderegg is unique in its own way, they all have a few things in common. Most are about the size and shape of a tennis ball, give or take quite a bit, and they usually have a brown or gray rind which is rough like a cauliflower and marked by a simple pattern of several more or less distinct ridges. Their insides are filled with some surprising combination of quartz crystals and gray agate which is usually marked with a pattern and often with splashes of brilliant color.

If a thunderegg is tenderly sawed in half, it is easy to see that the agate and quartz crystals fill an angular cavity that looks more or less star-shaped on the cut surface. If by pure luck the saw happens to section the thunderegg in just the right direction, the opposite sides of the cavity filling will show the outlines of a round bulge and depression about the size of a marble which would fit snugly into each other if they could be moved together. There is no way to tell from the outside where to cut to see these so relatively few sawed thundereggs show them.

Evidently thundereggs must begin as marble-sized lumps or holes within the volcanic ash which later expand to become angular cavities which then fill with agate and quartz crystals. It would be interesting to know what that original marble-sized lump or hole was, how and why it formed in the first place, and why it later expanded to make the angular hole in the rock. It would also be interesting to know why part of the silica filling is agate and the rest quartz crystals. And why do the rhyolite rinds of thundereggs get so solid that they resist weathering and stream transportation while the enclosing volcanic ash, which is also rhyolite, remains much softer? Somebody ought to figure all those things out.

For about 8 miles both north and south of Madras the highway crosses sedimentary rocks consisting essentially of reworked volcanic debris. Then it crosses another patch of John Day rhyolite ash deposits in the area immediately west of Smith Rock State Park. Smith Rock is also John Day rhyolite ash. The stretch of road from the area about 5 miles north of Terrebonne to Bend crosses basalt lava flows; those between Redmond and Bend are quite young.

Pilot Butte, the prominent conical mountain that overlooks Bend from the east is a basalt cinder cone still young enough to be almost unmarked by erosion but old enough to be cloaked in green vegetation. Estimating the ages of cinder cones by their appearance is a risky business because they always seem to be older than they look but this one can't be very old. It won't erupt again; cinder cones never do.

This basalt cast cooled around a burning tree as the surface of a lava flow dropped after having flooded higher. Lava Cast Forest south of Bend.

u.s. 97

bend —oregon 58

The route between Bend and Oregon 58 follows the eastern margin of the high Cascades through a series of fascinating landscapes. Every rock and mountain along the road is volcanic in one way or another and so are the landscapes except for a few stream valleys.

About 10 miles south of Bend, U.S. 97 passes the edge of an absolutely fresh flow of very ragged and blocky basalt at the base of Lava Butte, a cinder cone that rises about 500 feet immediately west of the highway. A narrow but nicely paved road, the "Phil Brogan trail," spirals to the top of the volcano where there is a small observatory building which contains an outstanding collection of accurately labelled volcanic rocks as well as panoramic paintings identifying the many Cascade volcanoes along the horizon.

Lava Butte is one of several very recent cinder cones and lava flows aligned along a fissure that extends about 20 miles northwest from the caldera of Newberry volcano. Eruptions seem to have worked their way northwestward along this line during the last 6000 years with Lava Butte being the most recent of the series. The next eruption will presumably occur several miles northwest of Lava Butte if the trend of the recent past continues into the future. This line of recent eruptions appears to be the westernmost continuation of the Brothers fault zone past Newberry volcano. It probably marks a very deep and actively moving break in the earth's crust.

Trees occasionally leave impressions of themselves cast in solid basalt although these are rare because molten basalt is so hot that it normally burns trees long before the lava is cool enough to retain impressions. Lava cast forests usually form in places where the first surge of basalt flooded deeply into a group of trees and then drained quickly away leaving the burning trees encased in a thin shell of

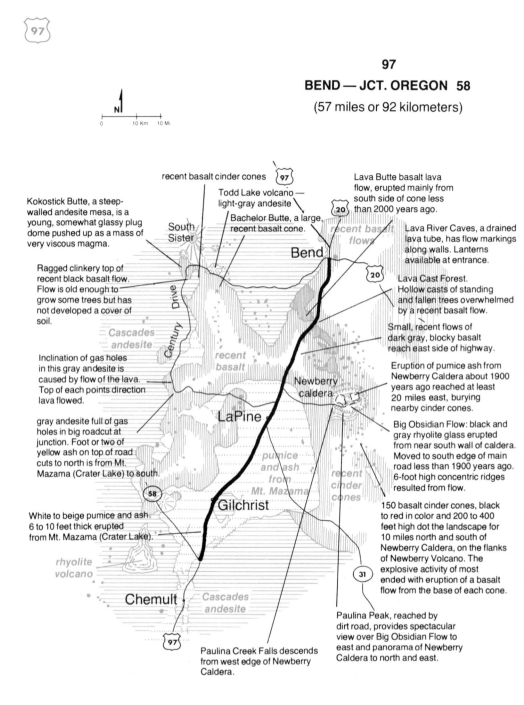

N

0   10 Km   10 Mi

recent basalt cinder cones 97

Todd Lake volcano —
light-gray andesite

Bachelor Butte, a large,
recent basalt cone.

South
Sister

Bend

Lava Butte basalt lava
flow, erupted mainly from
south side of cone less
than 2000 years ago.

20

recent basalt
flows

Lava River Caves, a drained
lava tube, has flow markings
along walls. Lanterns
available at entrance.

20

Kokostick Butte, a steep-
walled andesite mesa, is a
young, somewhat glassy plug
dome pushed up as a mass of
very viscous magma.

Lava Cast Forest.
Hollow casts of standing
and fallen trees overwhelmed
by a recent basalt flow.

Ragged clinkery top of
recent black basalt flow.
Flow is old enough to
grow some trees but has
not developed a cover of
soil.

Cascades
andesite

Century Drive

recent
basalt

Small, recent flows of
dark gray, blocky basalt
reach east side of highway.

Inclination of gas holes
in this gray andesite is
caused by flow of the lava.
Top of each points direction
lava flowed.

Newberry
caldera

Eruption of pumice ash from
Newberry Caldera about 1900
years ago reached at least
20 miles east, burying
nearby cinder cones.

gray andesite full of gas
holes in big roadcut at
junction. Foot or two of
yellow ash on top of road
cuts to north is from Mt.
Mazama (Crater Lake) to south.

LaPine

pumice
and ash
from
Mt. Mazama

recent
cinder
cones

Big Obsidian Flow: black and
gray rhyolite glass erupted
from near south wall of caldera.
Moved to south edge of main
road less than 1900 years ago.
6-foot high concentric ridges
resulted from flow.

58

Gilchrist

150 basalt cinder cones, black
to red in color and 200 to 400
feet high dot the landscape for
10 miles north and south of
Newberry Caldera, on the flanks
of Newberry Volcano. The
explosive activity of most
ended with eruption of a basalt
flow from the base of each cone.

White to beige pumice and ash
6 to 10 feet thick erupted
from Mt. Mazama (Crater Lake).

rhyolite
volcano

31

Chemult

Cascades
andesite

Paulina Peak, reached by
dirt road, provides spectacular
view over Big Obsidian Flow to
east and panorama of Newberry
Caldera to north and east.

97

Paulina Creek Falls descends
from west edge of Newberry
Caldera.

dripping basalt that solidified once it was exposed to the air. The tree casts reach up to the "high lava mark."

The Bend area has seen a lot of very recent volcanic activity, much of it in easily accessible places. Lava caves are always one of the main attractions in these areas. They form when the outer surface of a lava flow gets cool enough to solidify while the interior is still molten. The fluid magma within runs out from under the solid crust leaving the flow hollow inside. The molten rock lining the new cave often drips and runs for a while before it cools, making all sorts of weird little formations quite unlike those in limestone caves.

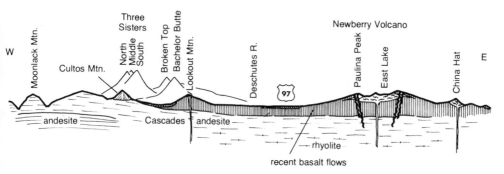

Section across the Cascades and through Newberry Volcano – just south of Bend.

Fantastic accumulations of ice sometimes fill lava caves making them much more beautiful and slightly more interesting than they would otherwise be. Ice accumulates where air circulation is restricted to more or less vertical openings, a common situation in lava caves. Cold air is denser than warm so new air can sink down into the cave only on days when the temperature outside is colder than that inside. Rock is a superb insulator so the cave tends to hold its temperature quite well and the ice survives from one season to another. Ice caves are not leftovers from the ice age or anything of the sort. Such tales are romantic nonsense. They form today wherever caves in fairly cold climates breath through vertical openings.

Between Bend and La Pine the road skirts the western flank of Newberry volcano, perhaps the largest, probably the most extraordinary, and certainly the least visible of all the volcanoes in the Oregon Cascades. Despite its enormous size, Newberry volcano doesn't make much of a landmark and isn't actually easy to see from this or any other highway because it is so broad and its flanks so gently sloping that it makes a very subtle outline against the sky. The only way to really enjoy and appreciate its unusual geology is to take the side road east from U.S. 97 to Paulina and East lakes which are right in the collapsed floor of the caldera and a wonderful place for a summer outing.

Newberry volcano is nearly 50 miles east of the main trend of the Cascade chain and has produced very little andesite. It began its career by erupting large volumes of basalt along with some rhyolite to build a broad dome, a very large shield volcano, approximately 25 miles in diameter. Then the entire top of the volcano collapsed to form a subsidence crater, called a caldera, that is about 8 miles in diameter. Volcanoes often do this as molten magma is withdrawn from beneath them to feed big eruptions.

After the caldera collapsed, Newberry volcano began to erupt rhyolite magmas in the form of ash, pumice and enormous obsidian lava flows. It built a small secondary pumice volcano right in the middle of the caldera depression and produced two huge obsidian flows which poured down the sides of the caldera walls toward its floor. One of these, along with the pumice cone, separates Paulina and East Lakes. Radiocarbon dates done on charred logs in the pumice show that the latest eruptions happened about 1900 years ago.

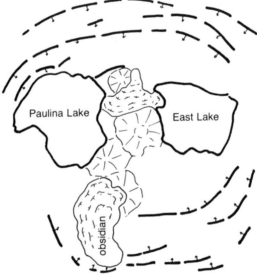

Generalized geologic map showing some of the volcanic rocks erupted since the caldera crater formed in the crest of Newberry volcano. The concentric pattern of dashed lines shows the system of faults along which the crater collapsed.

–adapted from Orebin, April, 1967

Although they could hardly look more different, pumice and obsidian are both glassy varieties of rhyolite, the light-colored volcanic rock that is comparable in composition to granite. Pumice is a light-colored rock so full of tiny bubbles that it really is a froth of glass, something like the sort of thing you might get by freezing soap suds. Obsidian, on the other hand, is shiny black glass with very few holes in it and about the same density as most other rocks. The black color is easy to explain; it is due to a very small amount of iron dissolved in the glass and staining it the way a few drops of ink might stain a glass full of water.

We know that rhyolite magmas are capable of holding quite a bit of water dissolved in them as steam. Evidently pumice forms when wet rhyolite magma reaches the surface and begins to boil off its steam. Rhyolite magma is very thick so the steam can't escape and the rock puffs up, as though it were bread dough, to make pumice.

Chemical analyses of obsidian show that it contains almost no water and is one of the driest rocks known. Evidently it forms from rhyolite magma that manages to reach the surface without picking up any water either when it melts or as it rises through the earth's crust. So the difference between pumice and obsidian seems to be mostly a matter of whether or not the rhyolite magma from which they both form happened to be wet or dry when it erupted.

Generations of geology students have learned that obsidian is a glass because the magma cooled so quickly that it didn't have time to crystallize. Whoever thought of that explanation obviously didn't know about the big obsidian flows in Newberry Caldera or the similar ones elsewhere in Oregon and other parts of the western states. There is no possible way that a lava flow the size of those in Newberry Caldera could cool quickly. Flows that size would probably take months to cool well below their freezing point. There must be some better explanation for the fact that obsidian is a glass; probably its lack of water prevents crystals from beginning to form.

The fact that Newberry volcano is perched right on the Brothers fault zone somehow seems very significant even though no one is quite sure just what it means. It has its twin in the Medicine Lake highlands of northern California a few miles south of the Oregon line. Both volcanoes are about the same distance east of the main trend of the Cascades; both began as broad shield volcanoes, both collapsed to form caldera craters, both are surrounded by dozens of small cinder cones, and both have recently erupted pumice and obsidian.

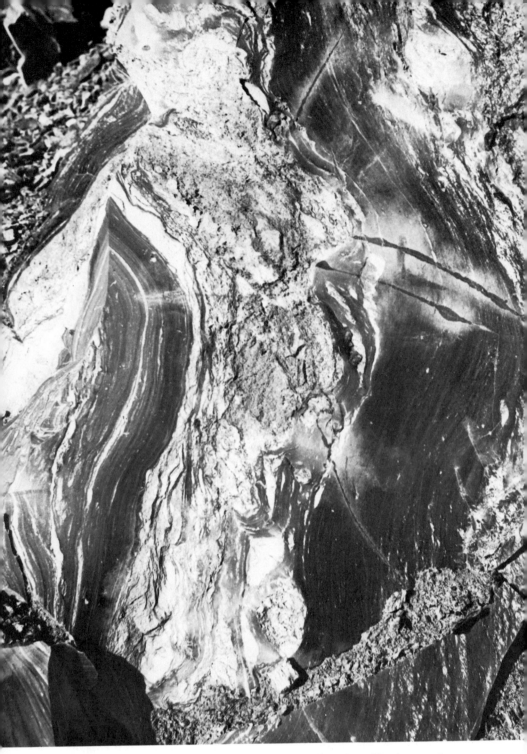

Streaky flow banding in black and gray obsidian from Newberry volcano.

u.s. 97

oregon 58 — california line

For approximately 35 miles, from the area about 3 miles south of the junction of highway 58 to that about 30 miles south, highway 97 crosses the thickest part of a vast, flat field of pumice erupted from Mount Mazama about 7000 years ago. Mazama had been a towering volcano comparable in size to Shasta until it destroyed itself in a great eruption that poured fiery avalanches of hot pumice over the nearby landscape and sent a cloud of volcanic ash as far northeast as central Montana and buried the area along the highway under a blanket of pumice that averages about 50 feet thick. The pumice field around the volcano is almost perfectly flat and quite variable in thickness because it filled low spots in the landscape that had existed before the eruption. All that remains of Mount Mazama is profiled against the skyline west of the pumice field.

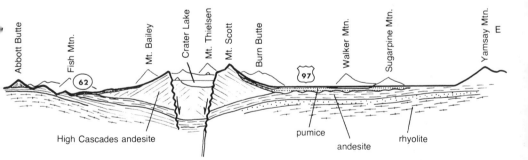

Section across the line of U.S. 97 between Chemult and Chiloquin showing how pumice and ash from Mount Mazama have buried the old landscape in this area.

The pumice deposit thins very rapidly southward in the vicinity of Solomon Butte, a conspicuous young volcano a few miles east of the road. In this area the highway crosses a series of fresh lava flows

97
JCT. OREG. 58 — KLAMATH FALLS & CALIFORNIA LINE
(80 miles or 129 kilometers)

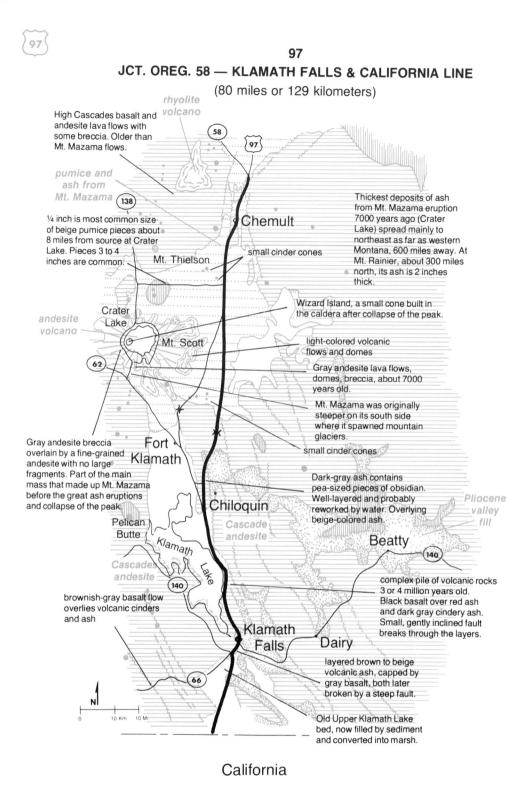

High Cascades basalt and andesite lava flows with some breccia. Older than Mt. Mazama flows.

rhyolite volcano

pumice and ash from Mt. Mazama

¼ inch is most common size of beige pumice pieces about 8 miles from source at Crater Lake. Pieces 3 to 4 inches are common.

Chemult

Mt. Thielson

small cinder cones

Thickest deposits of ash from Mt. Mazama eruption 7000 years ago (Crater Lake) spread mainly to northeast as far as western Montana, 600 miles away. At Mt. Rainier, about 300 miles north, its ash is 2 inches thick.

Wizard Island, a small cone built in the caldera after collapse of the peak.

andesite volcano

Crater Lake

Mt. Scott

light-colored volcanic flows and domes

Gray andesite lava flows, domes, breccia, about 7000 years old.

Mt. Mazama was originally steeper on its south side where it spawned mountain glaciers.

Gray andesite breccia overlain by a fine-grained andesite with no large fragments. Part of the main mass that made up Mt. Mazama before the great ash eruptions and collapse of the peak.

Fort Klamath

small cinder cones

Dark-gray ash contains pea-sized pieces of obsidian. Well-layered and probably reworked by water. Overlying beige-colored ash.

Pliocene valley fill

Chiloquin

Cascade andesite

Pelican Butte

Klamath Lake

Beatty

Cascade andesite

complex pile of volcanic rocks 3 or 4 million years old. Black basalt over red ash and dark gray cindery ash. Small, gently inclined fault breaks through the layers.

brownish-gray basalt flow overlies volcanic cinders and ash

Klamath Falls

Dairy

layered brown to beige volcanic ash, capped by gray basalt, both later broken by a steep fault.

N

0   10 Km   10 Mi

Old Upper Klamath Lake bed, now filled by sediment and converted into marsh.

California

erupted from Solomon Butte so thinly covered by Mazama ash and pumice that exposures of basalt are still visible in places.

About 7 miles north of Chiloquin where Mazama ash has thinned to about 3 feet, U.S. 97 crosses a steep grade that separates highlands to the north which are underlain by fairly young basalt lava flows, from the lush Klamath Lake lowlands to the south which are mostly underlain by much older lake deposits. Several spectacular roadcuts along the grade expose the lake deposits which consist of thin layers of black, basaltic, volcanic ash along with occasional beds of white diatomite, a lake sediment composed of the skeletons of microscopic algae.

Pumice beside U.S. 97 south of Chemult. The bigger chunks are about the size of walnuts.

Along most of the route between Chiloquin and Klamath Falls the road follows a narrow shelf between Klamath Lake to the west and high cliffs of very young volcanic rocks to the east. The cliffs are fault scarps created by recent crustal movements that created the basin now filled by Klamath Lake. Faults in this area trend generally northwesterly and have moved vertically, raising the hills and lowering the valleys. Many of them show signs of quite recent movement so it seems likely that these faults are still active and likely to move again causing earthquakes.

Klamath Falls is fortunate in this time of growing energy shortages to be right on top of an impressive deposit of natural steam. Years ago there were hot springs right in town and although these no longer flow, it is still possible to get hot water and steam from shallow wells nearly anywhere in town. Natural hot water heats the buildings of the Oregon Technical Institute as well as a number of private homes and there is plenty of dry steam to generate some electricity. Klamath Falls is one community that need not worry about freezing in the dark.

Obviously there must be very hot rock at shallow depth in the Klamath Falls area. This is hardly surprising since there are so many young volcanic rocks at the surface in the surrounding countryside. Evidently the old lake beds that underlie the surface of most of the Klamath Lake lowlands form a watertight lid over the hot rocks below trapping hot water and steam.

South of Klamath Falls U.S. 97 crosses Lower Klamath Lake, which is really mostly marshland, on a long causeway past an interesting system of drainage ditches. Another area of natural hot water and steam is in the Klamath Hills, visible about 8 miles east of the highway along the east side of Lower Klamath Lake. Wells in that area produce water too hot for cattle to drink so the ranchers have to cool it for several days in big storage tanks. That is the kind of problem that a few greenhouses might turn into a big profit.

u.s. 395

pendleton — john day

Most of the route between Pendleton and John Day crosses the basalt flows of the Columbia lava plateau; some parts of the road cross the gently rolling top of the plateau and others follow stream valleys eroded into it. The road also crosses outpost ridges of the Blue Mountains that rise like islands above the sea of basalt. These are composed of a variety of much older rocks that had formed and been eroded into hills long before the basalt eruptions began to build the plateau.

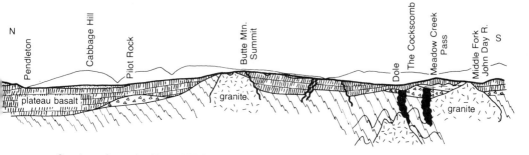

Section along the line of U.S. 395 between Pendleton and the Middle Fork of the John Day River – the northern half of the route between Pendleton and Mount Vernon.

Between Pendleton and Pilot Rock the top of the lava plateau is covered by coarse gravel deposited during Pliocene time, roughly between 3 and 10 million years ago. These gravels appear to fill old stream valleys that had existed before Pliocene time. The fact that the modern stream drainage cuts right across them shows that the modern streams began to erode their valleys after the Pliocene gravels were deposited.

At Nye the road finally gets out of the stream valley and onto the undulating surface of the lava plateau which it crosses to Battle Mountain, a long ridge of the Blue Mountains that was high enough

395

PENDLETON — JOHN DAY

(137 miles or 220 kilometers)

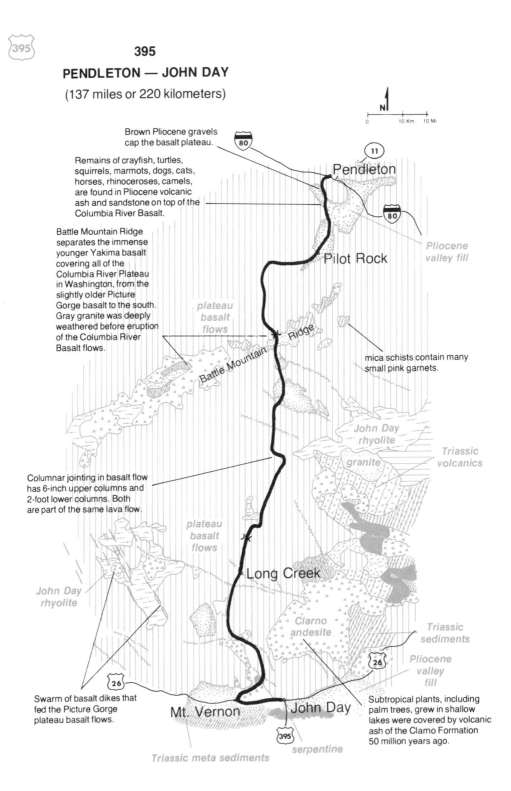

N

0   10 Km   10 Mi

Brown Pliocene gravels cap the basalt plateau.

Remains of crayfish, turtles, squirrels, marmots, dogs, cats, horses, rhinoceroses, camels, are found in Pliocene volcanic ash and sandstone on top of the Columbia River Basalt.

Battle Mountain Ridge separates the immense younger Yakima basalt covering all of the Columbia River Plateau in Washington, from the slightly older Picture Gorge basalt to the south. Gray granite was deeply weathered before eruption of the Columbia River Basalt flows.

Columnar jointing in basalt flow has 6-inch upper columns and 2-foot lower columns. Both are part of the same lava flow.

Swarm of basalt dikes that fed the Picture Gorge plateau basalt flows.

Pendleton

Pilot Rock

Pliocene valley fill

plateau basalt flows

Battle Mountain Ridge

mica schists contain many small pink garnets.

John Day rhyolite

granite

Triassic volcanics

plateau basalt flows

John Day rhyolite

Long Creek

Clarno andesite

Triassic sediments

Pliocene valley fill

Mt. Vernon   John Day

Subtropical plants, including palm trees, grew in shallow lakes were covered by volcanic ash of the Clarno Formation 50 million years ago.

serpentine

Triassic meta sediments

to keep its crest above the flood of basalt that covered this part of Oregon. Much of the ridge, which trends from southwest to northeast, is made of crumpled sedimentary rocks but these are intruded in places by bodies of granite and the road happens to cross one of those. Battle Mountain ridge was the barrier that separated flows of the Grande Ronde volcano to the north from those of the Picture Gorge volcano to the south.

A flow of Picture Gorge basalt buries a red soil developed on top of an older flow. Photographed beside U.S. 395 in the canyon near Dale.

There are some smaller and much less conspicuous outcrops of granite beside the road about halfway between Dale and Long Creek where it crosses another ridge of the Blue Mountains. Quite a few other knobs of older rocks protrude above the basalt in scattered low hills but these are the only two that the road actually crosses.

Between Long Creek and Mount Vernon the road crosses nearly a dozen faults, all trending approximately northwest to southeast, which have moved slivers of rock up and down, adding variety to both the landscape and the rocks.

For several miles along the creek valley immediately north of Mount Vernon, U.S. 395 passes large masses of serpentinite which are exposed in roadcuts and in big bald spots on the hillsides where grass will hardly grow and the dull greenish color of the bedrock shows through. Serpentinite forms in the oceanic crust and these masses are fragments of bedrock seafloor that got themselves scrambled into the mountains.

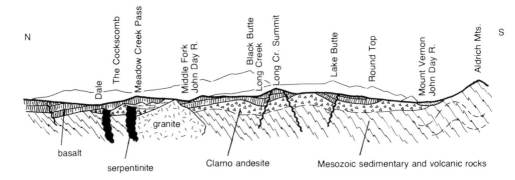

Section following the line of U.S. 395 along the southern half of the route between Pendleton and Mount Vernon.

Between Mount Vernon and John Day, U.S. 395 follows the valley of the John Day River with the Strawberry and Blue Mountains on the skyline to the east and the Aldrich Mountains on the south. Several roadcuts near Mount Vernon expose dark andesitic rubble which formed when a chain of volcanoes was active in this area about 45 million or so years ago. They belong to the Clarno Formation which makes up a large part of the Ochoco Mountains near Prineville.

232

u.s. 395

john day — burns

The drive between John Day and Burns passes some unusual rocks which tell us a great deal about the early origins of Oregon. The countryside is fairly well covered by trees and soil but there are enough outcrops to permit enjoyment of the geology.

Canyon City, just south of John Day, started with a boom as a gold mining town during the Civil War and continued in the mining business on a constantly declining scale until the early years of this century. Most of the production came early from placer deposits in the creek gravels. Several small lode deposits in the hills never amounted to very much.

For several miles south of town the road up the canyon passes large exposures of serpentinite which are easy to recognize by their dull greenish color. Specimens of serpentinite look greasy and feel almost like soap. Outcrops of it are full of slick surfaces with a high polish acquired as the weak rock slipped and flowed under pressure. Strange stuff; one of the oddest rocks.

Serpentinite forms when peridotite, the rock that makes most of the earth's interior, absorbs water. Its native habitat is beneath the layer of basalt lava flows on the seafloor and it shouldn't be sticking out of roadcuts in central Oregon unless there is some oceanic crust around. As it happens, Canyon Mountain, which is just east of the road, is a big slab of oceanic crust standing on edge but otherwise nearly intact. It was part of the seafloor about 200 million years ago until it somehow managed to get itself scrambled into the continent instead of sliding back into the inside of the earth, where it belongs, when the line of seafloor sinking was in this area some 150 million or so years ago.

395

JOHN DAY — BURNS

(70 miles or 133 kilometers)

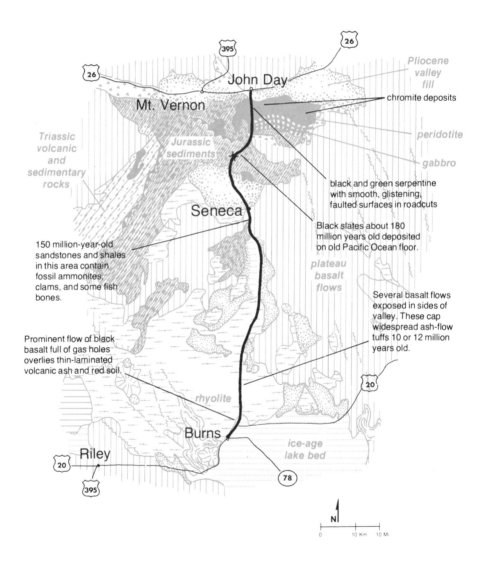

Pliocene valley fill

chromite deposits

peridotite

gabbro

Triassic volcanic and sedimentary rocks

Jurassic sediments

John Day

Mt. Vernon

black and green serpentine with smooth, glistening, faulted surfaces in roadcuts

Black slates about 180 million years old deposited on old Pacific Ocean floor.

Seneca

150 million-year-old sandstones and shales in this area contain fossil ammonites, clams, and some fish bones.

plateau basalt flows

Several basalt flows exposed in sides of valley. These cap widespread ash-flow tuffs 10 or 12 million years old.

Prominent flow of black basalt full of gas holes overlies thin-laminated volcanic ash and red soil.

rhyolite

Burns

ice-age lake bed

Riley

N

0   10 Km   10 Mi

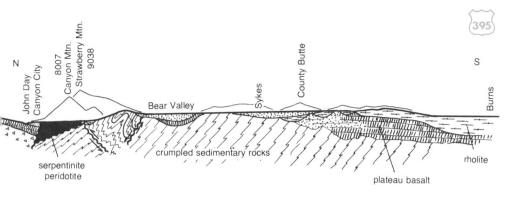

N — John Day — Canyon City — 8007 Canyon Mtn. — Strawberry Mtn. 9038 — Bear Valley — Sykes — County Butte — S — Burns

serpentinite peridotite — crumpled sedimentary rocks — plateau basalt — rholite

Section along the line of U.S. 395 between John Day and Burns.

Canyon Mountain is one of the finest slabs of oceanic crust exposed on a continent anywhere. Unfortunately there are so many faults near the road that the only way to see it clearly is to walk across the mountain along a line a few miles east of U.S. 395 and parallel to it. That may sound like a nuisance but it is actually an easy way to see a cross-section of the oceanic crust.

South of the serpentinite outcrops, south of the old slab of oceanic crust, the road passes through several miles of very slaty, black sedimentary rocks beautifully exposed in a series of nice clean road-cuts along a steep grade. These are old sediments that were deposits of sand and mud on the seafloor before they got scraped off onto the continent as it sank out from under them.

Slaty black mudstones in a roadcut about 15 miles south of John Day.

About 15 miles south of John Day, the road reaches the level of the top of the lava plateau and crosses a broad valley floored with Pliocene gravels washed into it several million years ago when the climate was much drier than it is now. Seneca is at the south side of the valley.

Just south of Seneca the road crosses a low ridge that stands slightly above the level of the lava plateau. There are more crumpled seafloor sediments in this ridge.

A heavy flow of black basalt covers beds of white volcanic ash beside the road a few miles north of Burns.

Silvies is near the north end of a broad upland valley which the road crosses for about 15 miles. There are no bedrock exposures in the valley floor because it is covered with gravels washed into it during Pliocene time and with deposits of light-colored volcanic ash.

All of the solid rocks between Silvies and Burns are volcanics. Their ages aren't certainly known but it seems a fair guess that they probably erupted during Miocene time, most likely about 15 or 20 million years ago. Most of the rocks conspicuously visible from the road are basalt lava flows which make strong ledges weathered to dark shades of brown. There are also a few big exposures of much paler and very rubbly andesite. Here and there along the way between Silvies and Burns are outcrops of pale volcanic ash in several pastel colors. The road comes into Burns down the valley of Poison Creek past beautiful big ledges of brownish black basalt resting on volcanic ash.

u.s. 395

burns — california line

The long route between Burns and Goose Lake on the California line threads the valleys of a mountainous desert in which the bedrock consists almost entirely of black basalt lava flows. There are some deposits of light-colored volcanic ash but these are much softer rocks than the basalt so they rarely form conspicuous outcrops. Sand dunes, alkali lakes, and shorelines of ice-age lakes add detail to the landscape and interest to the drive.

Volcanic eruptions between about 25 and 10 million years ago formed a lava plateau in south-central Oregon, burying all the older rocks. Then, much more recently, the area broke into blocks that moved up and down along faults to form the mountains and valleys we see today. That movement still continues and the present landscape consists essentially of fault-block mountains slightly carved by erosion and fault-block valleys floored by muddy sediments washed into them from the neighboring mountains. Younger volcanic rocks partially fill some of the valleys.

Because there is so little rainfall, connected systems of streams do not develop in deserts and the valleys normally remain undrained. When it rains, muddy runoff pours off the mountains and into the lowest part of the valley where it ponds and then dries up leaving the mud behind as the latest of the layers of sediment that floor the valley. So the valleys fill as the mountains erode.

Dissolved mineral matter also accumulates in desert valleys making the water ponded there alkaline so it glazes the ground with crusty white deposits as it evaporates. That is why the lakes that sparkle so deceptively in the desert valleys are not fertile oases but forbidding chemical sumps where evaporation concentrates the bitter residues of the desert. Most desert lakes are loaded with common salt and gypsum but a few contain valuable minerals such as borax.

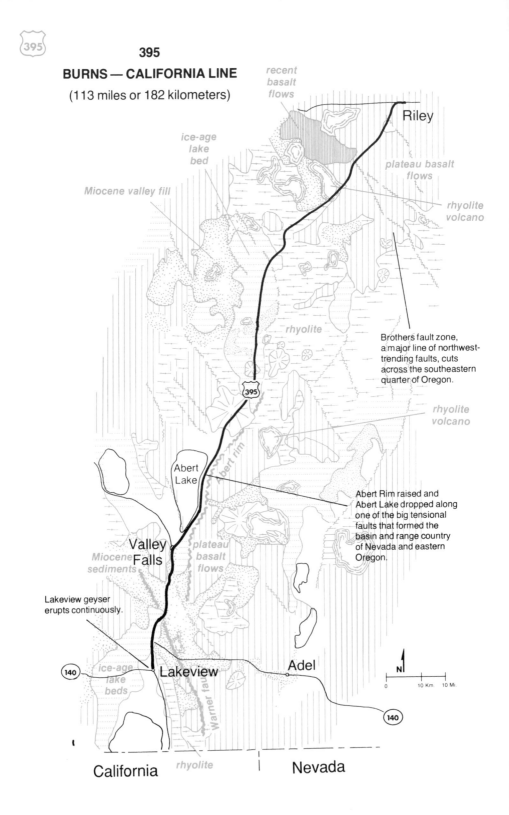

395

BURNS — CALIFORNIA LINE

(113 miles or 182 kilometers)

recent basalt flows

Riley

ice-age lake bed

plateau basalt flows

Miocene valley fill

rhyolite volcano

rhyolite

Brothers fault zone, a major line of northwest-trending faults, cuts across the southeastern quarter of Oregon.

395

rhyolite volcano

Abert Lake

Abert rim

Abert Rim raised and Abert Lake dropped along one of the big tensional faults that formed the basin and range country of Nevada and eastern Oregon.

Valley Falls

plateau basalt flows

Miocene sediments

Lakeview geyser erupts continuously.

Adel

N

0   10 Km.   10 Mi.

140

ice-age lake beds

Lakeview

Warner fault

140

California

rhyolite

Nevada

238

The ice ages brought the best refreshment desert valleys have known in millions of years. Those were very wet times when enough rain fell to fill them with large lakes that must have been reasonably fresh and to cloak the hills with green plants. But the ice ages didn't last very long from the geologic point of view, perhaps only a few tens of thousands of years, and the last one ended about 10,000 years ago. Now the desert is dry again and all that remains of the big lakes that brightened its valleys are old shorelines faintly grooved into the mountainsides and occasional deposits of fine silt in the floors of the valleys.

Burns is in the largest valley along the route of U.S. 395 and the big flat plain extending south and east of town toward Malheur lake is floored by lake-bed deposits. No old shorelines left by this lake are noticeable from the highway.

Between Riley Junction and the area about 8 miles south of Wagontiren U.S. 395 crosses at least 25 faults which run nearly at right angles to the road. They slice the lava plateau into hills that trend across the road and are softened just enough by erosion that the faults are hard to recognize from the ground. But in high altitude aerial photographs, or satellite photographs, the effects of erosion are invisible and the faults appear as sharp lines. This is the Brothers fault zone which extends from the area south of Jordan Valley north-westward to Newberry volcano just south of Bend. Everywhere south of the Brothers fault zone the lava plateau is broken into big fault-block mountain ranges and valleys; north of it the lava plateau is still relatively intact and unbroken by faulting. The difference is obvious in the scenery.

The bench along the base of this mountain overlooking Alkali Lake is the shoreline of a lake that filled the valley during the last ice age.

Between Wagontire and the California line, U.S. 395 passes through three large desert valleys separated by short stretches of hilly country. Alkali lake is just west of the road in the bottom of the northernmost of these. It is partly surrounded by a tract of sand dunes which the road crosses.

Dune beside U.S. 395 near Alkali Lake.

Abert Rim is one of the most spectacular fault scarps in the country. Erosion has hardly nicked it and very little debris has accumulated along its base to soften its sharpness. Such fresh-looking faults are convincing evidence that crustal movements in this part of Oregon are very recent and almost certainly still continuing.

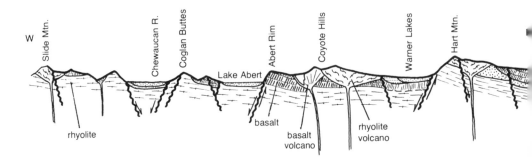

Section across the line of U.S. 395 at Abert Lake.

240

Between Valley Falls and the California line, U.S. 395 follows the base of the Warner Range which fills the skyline to the east. The range consists mostly of light-colored volcanic-ash deposits erupted during Miocene time, about 25 to 15 million years ago, and black flood basalts erupted in the last 10 million years. Movements along the Warner fault have raised the range to its present elevation within the last 10 million years. Along most of the route the fault is a mile or less east of the highway; in a number of places it makes such a steep slope that its location is fairly obvious.

Abert Rim, a fault scarp with Abert Lake at its toe. U.S. 395 passes between them.

Lakeview is between two areas of hot springs in which numerous hot water and steam wells have been drilled. Both of these areas are right along the highway, one about 2 miles north of town and the other about the same distance south. Hunter's Hot Springs, north of town, has been a popular resort for generations. A shallow well there produces boiling water by erupting like a small geyser about twice a minute.

Goose Lake floods the floor of a basin let down by movement along north-south trending faults within the last few million years. The lake used to be much larger and had an outlet into the Pit River in California. It has shrunk to a fraction of its former size within the memory of older people living in Lakeview, mostly because much of its former water supply is now diverted for irrigation. Now that the lake no longer has an outlet it is getting salty and it probably won't be too many years before its water is no longer fit for fish or anything else.

241

Basalt flow at Abert Rim full of large crystals of plagioclase feldspar. This is the "Steens basalt."

One of the things that geologists feel quite confident about is the fact that basalt lava flows are hopeless places to look for gold. But every such rule seems to have its exception and gold was indeed mined from basalt in the hills a few miles east of Pine Creek. Although some optimist called it the "High Grade" district, the deposits were actually small and production extremely modest. The district must have been found by someone who was either desperate or naive because no geologist or well informed prospector would look for gold in such rock.

oregon 3

washington line — enterprise

The entire route crosses basalt lava flows so there is never any doubt about the identity of the rocks along the road. This is one of the most interesting parts of the Columbia lava plateau and the rocks along this lonely road tell us more about its origin than those along any other road in the state. It is a fascinating drive.

The trip is also an adventure in motoring. Just north of the state line, the road winds down into the bottom of the Grande Ronde Canyon which is an impressive chasm by anyone's standards, crosses the river, and then corkscrews its way back up onto the plateau surface above. People who like switchbacks will love this drive.

All along the canyon walls there are beautiful roadcuts exposing lava flows which all seem to be quite thin; some are only a few feet thick. This is a striking contrast to the massively thick flows so familiar in most other parts of the Columbia Plateau. It seems very likely that the canyon cuts across the flank of the volcano that built the lava plateau and that these flows are thin because they spread out as they poured down its gentle slopes and were not ponded like those farther west.

Many of the roadcuts expose contacts between two flows and in nearly every case these are marked by rather well-developed red soils, the kind that form in the tropics. It takes a long time for thick soils to form on basalt so we can be sure that eruptions must have been fairly widely spaced. That old volcano was probably not a very alarming neighbor for thousands of years, but it did have its moments.

Numerous big dikes of basalt are exposed in the walls and floor of Grande Ronde Canyon. Some of those beside the road near the bottom

3

WASHINGTON LINE — ENTERPRISE

(43 miles or 69 kilometers)

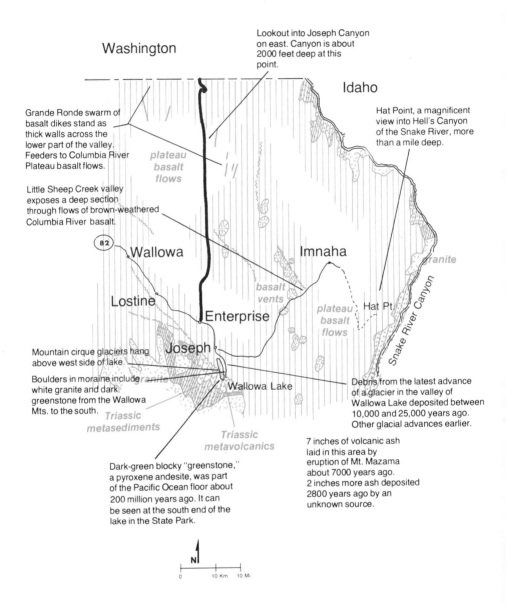

Washington

Idaho

Lookout into Joseph Canyon on east. Canyon is about 2000 feet deep at this point.

Hat Point, a magnificent view into Hell's Canyon of the Snake River, more than a mile deep.

Grande Ronde swarm of basalt dikes stand as thick walls across the lower part of the valley. Feeders to Columbia River Plateau basalt flows.

plateau basalt flows

Little Sheep Creek valley exposes a deep section through flows of brown-weathered Columbia River basalt.

82

Wallowa

Imnaha

granite

Lostine

basalt vents

Enterprise

plateau basalt flows Hat Pt.

Snake River Canyon

Mountain cirque glaciers hang above west side of lake.

Joseph

Boulders in moraine include white granite and dark greenstone from the Wallowa Mts. to the south.

Wallowa Lake

Debris from the latest advance of a glacier in the valley of Wallowa Lake deposited between 10,000 and 25,000 years ago. Other glacial advances earlier.

Triassic metasediments

Triassic metavolcanics

7 inches of volcanic ash laid in this area by eruption of Mt. Mazama about 7000 years ago. 2 inches more ash deposited 2800 years ago by an unknown source.

Dark-green blocky "greenstone," a pyroxene andesite, was part of the Pacific Ocean floor about 200 million years ago. It can be seen at the south end of the lake in the State Park.

N

0   10 Km.   10 Mi.

A big basalt dike stands like a wall beside the road in the bottom of Grande Ronde canyon. The shrinkage columns in dikes run crosswise making them look almost like stacked cordwood.

of the canyon stand up in relief, looking for all the world like huge ruined walls. The dikes in the canyon walls look like vertical seams cutting across the horizontal ledges that mark individual lava flows. This is one of the very few easily accessible places where large numbers of basalt dikes exist in the Columbia Plateau. They form a swarm that follows the eastern boundaries of Oregon and Washington and all the dikes trend approximately north-south.

Dikes are simply bodies of igneous rock that squirted into fractures in older rocks and solidified there. So a dike, being a fracture filling, is a thin, straight body of igneous rock shaped like a big board. All we normally see is one edge. It seems quite certain that the dikes in the Grande Ronde Canyon formed as basalt magma flowed upward through fractures to pour out on the surface as lava flows. The last bit of magma that didn't make it to the surface froze in the fracture to become a dike. So what we see in the dikes in the Grande Ronde canyon is the internal plumbing of the old volcano that created the Columbia lava plateau.

South of the Grande Ronde Canyon the road stays on the high plateau surface all the way to Enterprise. The only adventure is the view down into Joseph Canyon, another impressive gash eroded into the lava plateau. The walls of the canyon are a long flight of ledges, like giant steps, each marking one of the flows of basalt that built the plateau. The youngest of these flows erupted about 15 million years ago so erosion must have carved the canyon since then.

Between Joseph Canyon and Enterprise, the road follows the plateau surface through beautiful groves of trees with hardly a rock in sight. The main geologic interest here is out of sight, a series of small buttes hidden in the woods a few miles off to the east and arranged in a line roughly parallel to the road. Unpaved side roads lead directly to several of them. These little buttes are basalt cinder cones, old volcanoes. Radioactive ages recently obtained on rocks from some of them show that they are about 15 million years old, the same age as many eruptions on the Columbia Plateau. Evidently these are the actual volcanic vents from which some of the lava flows poured and they show that the plateau flows did indeed come from this area and that they were fed by the dikes exposed a few miles north in Joseph and Grand Ronde canyons.

This is the highest part of the Columbia Plateau, the only place in the largest part of the plateau known to contain basalt dikes, and the only place in which volcanic vents have been recognized. It seems extremely likely that this area is actually the crest of the old volcano that built almost the entire Columbia Plateau, certainly one of the largest volcanoes ever to have existed on the earth. The pieces of the Columbia Plateau puzzle fit better along Oregon 3 than anywhere else.

Horizontal ledges of basalt outcropping in the canyon walls near Imnaha.

enterprise—imnaha—hat point

The secondary road from Enterprise to Hat Point is the only way for road travellers to get a good view of the Hell's Canyon of the Snake River which is as deep and impressive as the Grand Canyon. Hell's Canyon is in wild and nearly roadless country so very few people have seen it and the view from Hat Point is a rare treat. You have to walk to get to any other rim viewpoint on either side.

A few miles east of Joseph the road meets Little Sheep Creek which it follows downstream to the community of Imnaha deep in the canyon of the Imnaha River. From Imnaha the road is unpaved and winds up the steep canyon wall, back on to the plateau surface which it follows to Hat Point. This is a rough road that provides more adventure than most people care for. Bedrock along the way, for those

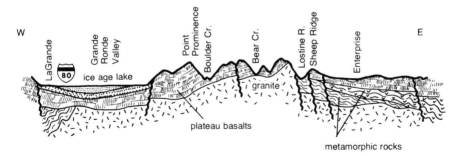

Section east from Joseph across the Hell's Canyon country.

who may have time to look at it while replacing the oil pan, is all flood basalt flows.

The view from Hat Point can neither be photographed nor described. Hell's Canyon here is slightly more than a mile deep and approximately 10 miles wide, fully as deep as the Grand Canyon and considerably narrower. This part of Oregon gets some rain so the canyon walls are thinly mantled by soil which shelters beneath a cloak of grass and a scattering cover of bushes and small trees. The dominant color scheme is in shades of green with the color of the bedrock showing through as an undertone.

Horizontal ledges of plateau basalt form bold outcrops in the upper part of the canyon walls, their dark color and tendency to break into vertical columns makes them easy to recognize. Beneath the basalt is a thick section of light-colored rock in which very little structure or pattern is visible from Hat Point. These are the so-called Seven Devil's sequence, a thick and confusing assemblage of crumpled volcanic and sedimentary rocks first laid down on the floor of the Pacific Ocean 200 million or more years ago, and then scraped off against the edge of the continent, then in westernmost Idaho, about 150 million years ago. These are clearly similar to the rocks in the Blue and Wallowa Mountains and to some in western Idaho. The Snake River has cut through the basalts, opening a geologic window through the lava plateau into the rocks beneath.

One of the many surprising things about Hell's Canyon is the fact that the surface of the plateau in which it is cut is highest at the canyon rim. The plateau surface slopes up to the canyon rim, not down towards it as we might expect. Many other large canyons are like this and the explanation is as curious as the phenomenon — it depends upon the fact that the earth's crust is floating on the interior.

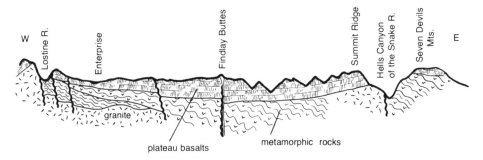

Section from Enterprise to Hell's Canyon showing how the plateau basalt covers the older rocks of the Blue Mountains.

248

Because the earth's crust is floating, it sinks when a load is placed on it and rises when one is removed, as though it were an air mattress floating on a lake. For example, loading the crust by filling the reservoir behind a dam will cause it to sink under the weight of the water to an extent that is accurately predictable in theory and easily measurable by ordinary surveying methods. Likewise, removing a load from the earth's crust will cause it to float upward to compensate for the weight of the material removed. Erosion of a large canyon, such as Hell's Canyon, removes a large load, causing the area of the canyon to rise. Erosion is not removing material from the nearly flat plateau surface away from the canyon rim. So as erosion carves Hell's Canyon deeper, its rim rises steadily higher and we see that under the right circumstances erosion can actually make mountains higher instead of lower.

The Seven Devil's Mountains which form the skyline southeast of Hat Point are capped by basalt lava flows which are higher than those in any other part of the Columbia Plateau. Obviously, they are uplifted. At least part of that uplift, if not all of it, is certainly due to the fact that the Seven Devil's Mountains are between Hell's Canyon and the Salmon River Canyon in Idaho and are floating upward in response to erosional unloading of the crust on both sides.

The glacial moraine right in the center of the picture is the dam that impounds Wallowa Lake.

oregon 82

la grande — joseph — wallowa lake

The road between La Grande and Enterprise skirts a northern spur of the Wallowa Mountains following a route across the thin edge of the plateau basalts near where they lap onto the older rocks to the south. All the rocks along the road are basalt lava flows. Bedrock in the high peaks of the Wallowa Mountains on the skyline east of LaGrande and south of Enterprise is mostly granite.

Between LaGrande and Minam the road gently rises and falls as it follows the softly undulating upper surface of the lava plateau. Flood basalts in this region erupted at about 15 million years ago and are now covered by deep and fertile soils that effectively hide the bedrock and support excellent crops. The widely scattered outcrops and road-cuts expose basalt weathered to a brownish-black color. They are especially well exposed along the long grade between the basalt plateau and the canyon of the Wallowa River at Minam.

82

LAGRANDE — ENTERPRISE

(65 miles or 105 kilometers)

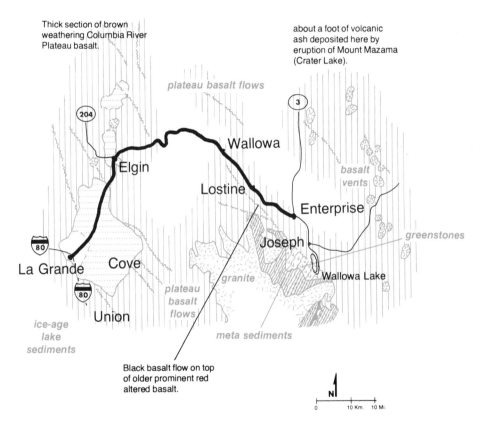

Thick section of brown weathering Columbia River Plateau basalt.

about a foot of volcanic ash deposited here by eruption of Mount Mazama (Crater Lake).

plateau basalt flows

Wallowa

Elgin

Lostine

basalt vents

Enterprise

greenstones

Joseph

La Grande

Cove

granite

Wallowa Lake

ice-age lake sediments

Union

plateau basalt flows

meta sediments

Black basalt flow on top of older prominent red altered basalt.

N

0   10 Km.   10 Mi.

Boulders of granite in glacial till, a roadcut in the moraine that impounds Wallowa Lake.

Between Minam and Wallowa, the road follows the Wallowa River through a narrow canyon cut into the basalt flows. East of Wallowa the canyon is wider and the road passes through open valleys with good views up long, gentle slopes to the dramatic profile of the Wallowa Mountains against the southern horizon.

An unusually interesting roadcut just over 2 miles east of Lostine exposes a thick flow of black basalt on top of an older flow capped by red soil. Evidently the red basalt was exposed at the surface for a long time and deeply weathered before a later eruption covered it with the upper lava flow. Outcrops like these show that the lava plateau was built by a series of eruptions spaced over long time intervals, not by a catastrophically rapid sequence of volcanic outbursts.

Between Enterprise and Joseph the road crosses a broad glacial outwash plain that rises very gently to a crest just south of Joseph, at the glacial moraine that impounds Wallowa Lake. The entire plain is covered by debris that washed out of that one glacier as it melted.

The north end of Wallowa Lake is neatly enclosed within a terminal moraine that forms a hairpin-shaped ridge outlining the edge of a huge ice-age glacier. The valley south of the lake is gouged out into a broad trough by glacial erosion. It is a magnificent scene. There were a lot of glaciers in the Wallowa Mountains during the last ice age which carved the mountains into the ragged and craggy mass of peaks we see today. Had it not been for those ice-age glaciers, the Wallowas would be a mass of gently rounded mountains with broadly convex tops instead of the sharp peaks we see today.

oregon 31

lapine (u.s. 97) - valley falls (u.s. 395)

The lonely road between LaPine and Valley Falls passes through beautiful scenery and interesting rocks along its entire route. The fact that every rock along the way is volcanic does not imply a lack of variety. Some of the most extraordinary rocks and landscapes in Oregon are along this road.

LaPine is at the southwestern edge of Newberry volcano, an enormous pile of basalt very broad in proportion to its height. Newberry's forested mass forms an imposing but quite unemphatic outline against the northern horizon wherever it is visible through the trees along the way between LaPine and the Fort Rock area.

The entire Fort Rock area is a garden of volcanic oddities which owe their strangeness to a shallow lake that flooded this valley during the last ice age. Basalt magma rising into the watery muds beneath the lake generated steam that powered violent volcanic outbursts. Dry magmas always erupt without undue commotion; it is only steam that inspires volcanoes to violence. Molten basalt is normally dry but its very high temperature of more than 1200° centigrade makes it perfectly capable of generating enough intensely live steam to add real emphasis to an eruption.

Fort Rock is a peculiar volcano that looks from the air like a giant doughnut with most of one side munched off. It was originally a complete ring of volcanic ash created when basalt magma got into the wet muds of the lake bottom and powered a jet of steam that blew molten basalt into the air in a cloud of tiny shreds. The eruption must have looked at night like an enormous geyser, blasting a glowing fountain of red hot magma into the air. The bits of basalt settled around the vent to make a ring of volcanic ash set as an island in the shallow waters of the lake. Then waves eroded the outside of the ring

31
LAPINE — VALLEY FALLS
(120 miles or 193 kilometers)

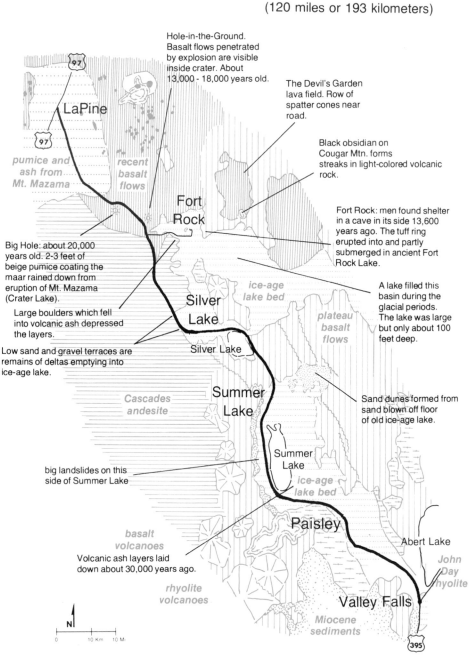

Hole-in-the-Ground. Basalt flows penetrated by explosion are visible inside crater. About 13,000 - 18,000 years old.

The Devil's Garden lava field. Row of spatter cones near road.

Black obsidian on Cougar Mtn. forms streaks in light-colored volcanic rock.

LaPine

pumice and ash from Mt. Mazama

recent basalt flows

Fort Rock: men found shelter in a cave in its side 13,600 years ago. The tuff ring erupted into and partly submerged in ancient Fort Rock Lake.

Fort Rock

Big Hole: about 20,000 years old. 2-3 feet of beige pumice coating the maar rained down from eruption of Mt. Mazama (Crater Lake).

A lake filled this basin during the glacial periods. The lake was large but only about 100 feet deep.

ice-age lake bed

Silver Lake

plateau basalt flows

Large boulders which fell into volcanic ash depressed the layers.

Low sand and gravel terraces are remains of deltas emptying into ice-age lake.

Silver Lake

Summer Lake

Cascades andesite

Sand dunes formed from sand blown off floor of old ice-age lake.

Summer Lake

big landslides on this side of Summer Lake

ice-age lake bed

Paisley

basalt volcanoes

Abert Lake

Volcanic ash layers laid down about 30,000 years ago.

John Day rhyolite

rhyolite volcanoes

N

0   10 Km   10 Mi

Valley Falls

Miocene sediments

and cut the steep cliffs that make Fort Rock look almost like a ruined castle. Ash layers exposed in the cliffs clearly reveal the internal structure of the volcano. It was wave erosion that bit the hole out of the south side of the ash ring.

It began as a volcanic island in a lake. Now Fort Rock's wave cut cliffs rise like battlements above the prairie. One of Oregon's most interesting landmarks.

Flat Top, a butte about 7 miles north of Fort Rock, is basically the same sort of thing although it looks quite different. There basalt magma rose quietly into the ash ring during the last stage of the eruption, filling it as though it were a soup bowl and overflowing the northwest rim. Evidently the growing ash ring made an effective dam that stopped lake water from flowing into the vent and thus cut off the steam jet.

Hole-in-the-Ground, just northeast of Oregon 31 a few miles west of Fort Rock, and Big Hole, on the opposite side of the road about 6 miles farther west, are both open craters created by steam explosions. Both are about a mile in diameter and several hundred feet deep. Both are surrounded by low rims of debris blown from the crater now covered by a coating of light-colored ash erupted from Newberry volcano. The blasts of steam that blew these holes into the landscape carried very little magma with them so Hole-in-the-Ground and Big Hole developed as open craters instead of ash rings.

Moffitt Butte, immediately east of Oregon 31 about halfway between LaPine and the Fort Rock turnoff is another ash ring and so is Table Mountain about 6 miles directly east of Fort Rock. There are several others in the vicinity.

The county road that goes east and then north from Fort Rock passes through the Devil's Garden volcanic field which is shown as lava beds on most highway maps. There was no lake here and very little steam so the action was much quieter. Eruptions in the Devil's garden created a number of small basalt cinder cones and extensive lava flows all of which are nearly bare of trees and look almost perfectly fresh. It would be easy to imagine that this activity might be no more than a few hundred years old. There are bits of light-colored pumice littering the surface that must have come from the eruption of Newberry volcano about 1900 years ago. So we can be sure that these fresh lava flows are at least that old, perhaps considerably older.

It is always sobering to look at such fresh lava surfaces, consider their age, and then reflect upon the slowness of the processes that break fresh rock down into fertile soil. Obviously, soil is not a renewable resource within any kind of humanly meaningful time span, even though we know that it forms constantly.

The relatively recent volcanic activity from Fort Rock northward is in the area of the Brothers fault zone that cuts across from southeastern Oregon to Newberry volcano. South of Fort Rock the landscape changes as it does everywhere south of the Brothers fault zone. The country opens out into a series of high fault block mountains separated by expansive valleys which are fault-block basins. Oregon 31 follows a route through the Silver and Summer Lake valleys past mountains composed entirely of volcanic rocks, most of them basalt lava flows erupted during Pliocene time, several million years ago. There are excellent exposures of these basalts complete with nice columnar jointing in Picture Rock Pass between Silver and Summer lakes. Ancient Indian artwork on some of the rocks gives the pass its name.

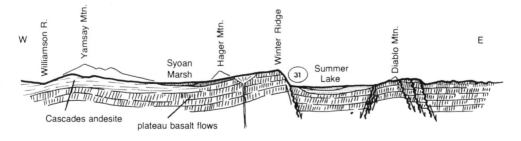

Section across the line of Oregon 31 at Summer Lake.

257

The big bank of gravel with the level top at the foot of Tucker Hill near Paisley is the shoreline of a lake that filled this valley during the last ice age.

Summer Lake floods the floor of a broad valley formed as a block of the earth's crust dropped along faults during the past few million years. Rainfall in this semi-arid region is just sufficient to keep part of the valley floor shallowly flooded but not enough to establish stream flow that would cut a spillway and drain the lake.

The road skirts the western shore of the lake, closely following the line of the fault that let the Summer Lake valley down and raised the mountains to the west. The hills raised by the fault west of the road consist of a thick sequence of rather weak volcanic-ash deposits twice the age of the massive basalt lava flows which cap them. The ash beds are too weak to bear the weight of the basalt and collapsed in a series of giant landslides that slumped toward the lake. The road winds along the toes of the slides, passing rough and broken hillsides, the kind of landscape that large landslides always seem to produce.

About halfway between Paisley and Valley Falls the road follows the Chewaucan river past the north end of Tucker Hill, a prominent landmark full of geologic interest. Its lower third is covered by a huge bench of gravel with a perfectly horizontal upper surface. This is an old beach that formed during the last ice age when this entire valley filled with water to make a huge lake and Tucker Hill was an island. Obviously the climate was much wetter then than it has been since.

You can't tell it from the road, but much of Tucker Hill is made of an interesting and useful kind of volcanic rock called perlite. This is an unusual variety of volcanic glass that has the startling property of puffing up like popcorn when it is heated. Expanded perlite is useful for several purposes, the most important of which is making lightweight concrete. Most people are more familiar with it as the little white kernels of light rock speckled through commercial potting soil. Tucker Hill is only one of many perlite deposits in this part of Oregon.

oregon 78

burns — burns junction (u.s. 95)

The route crosses the lava plateau of southeastern Oregon through a landscape broken into blocks by movement along countless faults during the past few million years. Every rock along the way is volcanic, the dark ones are all basalt and the light ones rhyolite.

Between Burns and Princeton the road crosses the flat floor of the Harney Valley. There are no outcrops but the flatness of the valley floor suggests that it is underlain by old lake sediments.

About midway between Princeton and Burns Junction, Oregon 78 crosses the northernmost tip of the Steens Mountain fault block. It isn't much of a mountain here but it rises to the south to become the highest mountain block in southeastern Oregon and the only one that held glaciers during the last ice age.

An excellent gravelled county road goes from Oregon 78 to Denio by way of the Alvord Valley and some of the most interesting rocks in southeastern Oregon. The road follows the eastern base of the Steens and Pueblo Mountain blocks, tracing the course of their bounding fault all the way. The Alvord Valley east of the road is a block that dropped along faults at the same time that the Steens and Pueblo blocks were rising.

There are numerous hot springs near the road all along the base of Steens Mountain and several in the floor of the Alvord Valley. They all seem to rise along faults. One of these is Hot Lake, a small lake with a large hot spring in its center about 2 miles south of Alvord Lake. Hot Lake is full of borax which supported an unusual mining operation between 1898 and 1907. Chinese laborers scraped up the white, salty crusts that coated the ground around the lake and then extracted the borax in big boilers fired by sagebrush. The operation

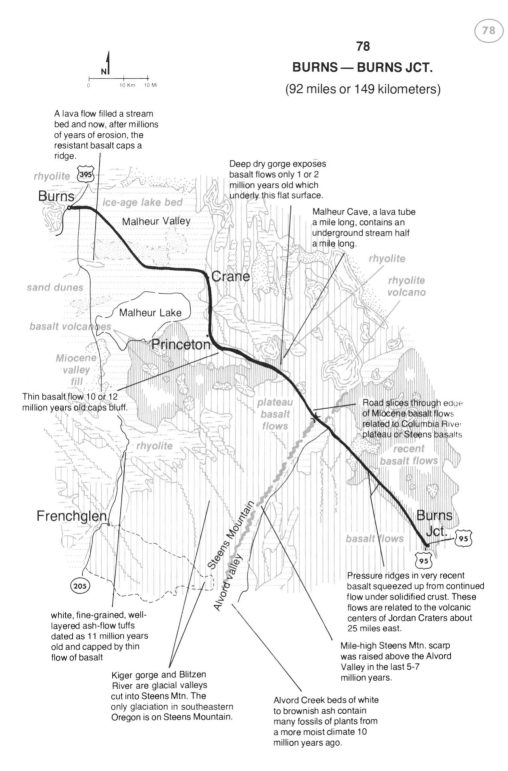

A lava flow filled a stream bed and now, after millions of years of erosion, the resistant basalt caps a ridge.

rhyolite (395)

Burns

ice-age lake bed

Malheur Valley

Deep dry gorge exposes basalt flows only 1 or 2 million years old which underly this flat surface.

Malheur Cave, a lava tube a mile long, contains an underground stream half a mile long.

rhyolite

sand dunes

Crane

rhyolite volcano

basalt volcanoes

Malheur Lake

Miocene valley fill

Princeton

Thin basalt flow 10 or 12 million years old caps bluff.

plateau basalt flows

Road slices through edge of Miocene basalt flows related to Columbia River plateau or Steens basalts

rhyolite

recent basalt flows

Frenchglen

Steens Mountain

Burns Jct.

basalt flows

(95)

(95)

(205)

Alvord Valley

Pressure ridges in very recent basalt squeezed up from continued flow under solidified crust. These flows are related to the volcanic centers of Jordan Craters about 25 miles east.

white, fine-grained, well-layered ash-flow tuffs dated as 11 million years old and capped by thin flow of basalt

Mile-high Steens Mtn. scarp was raised above the Alvord Valley in the last 5-7 million years.

Kiger gorge and Blitzen River are glacial valleys cut into Steens Mtn. The only glaciation in southeastern Oregon is on Steens Mountain.

Alvord Creek beds of white to brownish ash contain many fossils of plants from a more moist climate 10 million years ago.

An unusual curving fracture pattern in basalt beside the road about midway between Princeton and Burns Junction.

closed when the best salt deposits and the sagebrush both ran out at about the same time — this may well be the only known example of economic hardship caused by a shortage of sagebrush. Hardly anything remains of the mine today except some rusting vats off in the valley east of the road about 6 miles north of Fields.

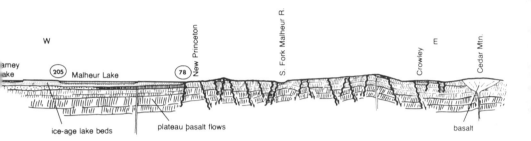

Section across the line of highway 78 near Princeton.

The eastern face of Steens Mountain contains a thick section of brown andesites erupted during Miocene time, probably between 20 and 25 million years ago. Farther south, between Pueblo and Denio, the same fault brings up crystalline rocks similar to some of those in the Klamath and Blue Mountains. Big boulders of gray granite and various other rocks are scattered along the road. This is the only place in southeastern Oregon where such rocks come to the surface. Their existence below the surface almost certainly explains why the lava plateau in this region contains so much rhyolite.

The section of Oregon 78 between Princeton and Burns Junction also crosses the Brothers fault zone which is poorly expressed in this area because it interferes with the Steens Mountain faults. But it exists just the same and has spawned some fairly recent volcanic activity. So some of the basalt flows near the road both north and south of the Alvord Valley road are much younger than the Miocene basalts that cover most of southeastern Oregon away from the Brothers fault zone.

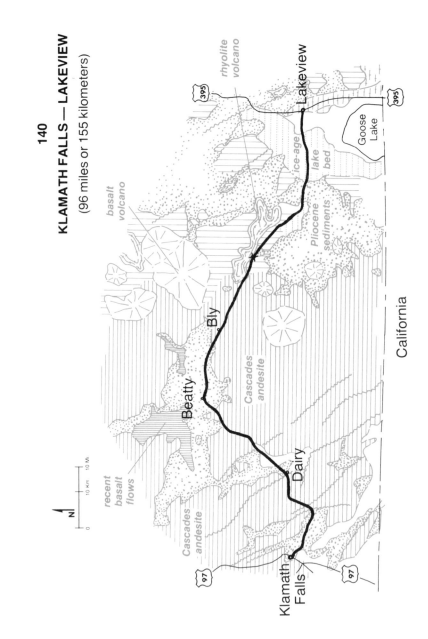

140

140

KLAMATH FALLS — LAKEVIEW

(96 miles or 155 kilometers)

oregon 140

klamath falls — lakeview

Nearly all the rocks along the route between Klamath Falls and Lakeview are volcanic. They formed over a period of several million years in a variety of ways so the details of their story are complicated. Nevertheless, it is fairly easy to sort them out and see the geologic history of the area expressed in the landscape along the road.

During Pliocene time, about 5 million or so years ago, large lakes flooded considerable areas of south-central Oregon. Their beds are still fairly flat today. Sediments that accumulated in them lie beneath most of the gently rolling low country around Klamath Falls and form the floors of the broad valleys the road follows most of the way between Klamath Falls and Bly.

The Pliocene lake sediments are soft and erode easily so they rarely form conspicuous outcrops. But they do show up in occasional road-cuts as thinly- layered brown and white sediments. Lake deposits typically accumulate in very thin layers, a fact that helps make them fairly easy to recognize. The brown layers consist mostly of volcanic ash altered by contact with water. Some of the light layers also contain volcanic ash but the starkly white ones are diatomite, an unusual sediment composed of the skeletons of microscopic algae called diatoms.

Lakes in volcanic areas usually support a lush growth of diatoms because their waters are rich in dissolved silica which functions as a fertilizer. Klamath lake, for example, turns into a rich, green diatom soup every summer. A high-powered microscope reveals diatom skeletons as beautifully delicate laceworks of fine silica needles far more fanciful and geometrically complex than any snowflake. Diatomite deposits are composed of millions of these intricately delicate skeletons that make a fluffy, lightweight material useful for all sorts of purposes ranging from fireproof safe insulation to filtering beer.

Most of the route between Klamath Falls and Bly crosses the Pliocene lake beds, passing through a succession of broad valleys separated by hills composed of volcanic rocks erupted on top of them. Anything that looks from this stretch of road as though it might be a volcano almost certainly is one.

Bug Butte seen looking north from Oregon 140 at its junction with the Sprague River Road is an andesite dome.

Most of the rocks exposed between Bly and Drews reservoir are basalt lava flows, some of which are older than the Pliocene lake beds and others younger. It is hard to tell which is which because basalts tend to look about the same regardless of their age.

Quartz Mountain, about halfway between Bly and Drews reservoir, is a rhyolite volcano, one of several north of the road in this general area. As often happens with this type of volcano, Quartz Mountain blew steam for many years after it finished its last eruption and stewed its rocks into a featureless mass of pale, opalized rhyolite. Not the kind of attractive opal that people make into gems, this is the ugly kind that often contains deposits of ore minerals.

There is mercury here and there in the opal in Quartz Mountain and the nearby hills, and several mines have worked it at various times over the years. About 10 miles east of Quartz Mountain there are several uranium deposits in another patch of altered rhyolites and these have also been worked from time to time over the years. Both mercury and uranium are fairly widespread wherever light-colored volcanic rocks occur in the hills between Quartz Mountain and Paisley.

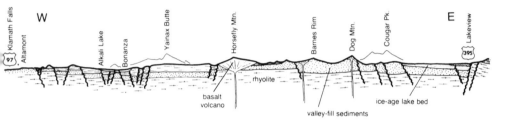

Section south of the line of the road between Klamath Falls and Lakeview.

Immediately east of Drews reservoir the road crosses about 5 miles of hilly country underlain mostly by volcanic-ash deposits older than the lake beds and basalt lava flows farther west. These volcanic rocks probably lie beneath much of the surrounding region to a depth of many thousands of feet but in most places they are buried beneath younger deposits.

Just 5 miles east of Drews reservoir the road leaves the hills and crosses onto a high bench of old lake deposits left by Goose Lake thousands of years ago when it stood much higher than it does now. Then it crosses a few miles of low lake bed sediments only a few feet above the modern water level and finally enters Lakeview.

glossary

Agate. A variety of quartz distinguished by its extremely fine grain size and bright colors. Agates may occur in almost any kind of rock but are especially common in volcanics.

Andesite. A common volcanic rock intermediate in composition between basalt and rhyolite. Andesite comes in various shades of gray, brown or green and commonly occurs as lava flows, ash deposits and accumulations of angular debris.

Basalt. The commonest volcanic rock and surely the most abundant rock in Oregon. Basalt is very fine-grained, has a smooth texture, and is quite black if fresh. Weathered or altered basalt may be greenish black or various rusty shades of brown, occasionally even brick red. Many specimens are full of gas bubbles.

Bauxite. A type of laterite soil that is very rich in aluminum and poor in iron. The best bauxites are nearly white but most of those in Oregon contain enough iron to color them red.

Caldera. A large, basin-shaped crater formed by collapse of a volcano during an eruption. Crater Lake is an excellent example.

Chromite. A mineral composed of chromium oxide. It is heavy and black and the only mineral source of chromium. Chromite always occurs in peridotite or serpentinite.

Cinder Cone. A small basalt volcano that erupts a conical pile of bubbly fragments and then produces one or two lava flows which emerge from the base of the cone.

Cretaceous. The interval of time between 135 and 70 million years ago.

Crust. The rigid outer part of the earth extending down to a depth of about 60 miles.

Dike. A body of igneous rock that formed when magma squirted into a fracture andcooled there. Most dikes are shaped like big slabs standing nearly vertically within the enclosing rock.

Diorite. A coarsely granular rock composed of milky crystals of feldspar and abundant grains of black hornblende or mica. It somewhat resembles granite except for being much darker and lacking quartz. Like granite, it forms when molten magma cools deep within the earth's crust.

Eocene. The period of time between about 60 and 40 million years ago.

Fault. A fracture in the earth's crust the opposite sides of which have shifted past each other.

Feldspar. An extremely common and abundant family of minerals most of which are rather milky looking. In light-colored rocks the feldspars are commonly pink or white; in dark-colored rocks they are usually either greenish or white.

Gabbro. A coarsely granular rock composed of greenish white feldspar and black pyroxene. It is usually very dark in color.

Granite. A granular rock composed of crystals of glassy-looking quartz, milky feldspar, and black hornblende or biotite. It forms when andesite magma cools very slowly beneath the surface.

Greenstone. Volcanic rocks that have been recrystallized at high temperature and pressure (metamorphosed). Their bright green color is both startling and distinctive.

Igneous Rock. A rock formed by cooling of a molten magma either on the surface after it has erupted from a volcano or at depth within the crust of the earth.

Jurassic. The geologic period that began about 180 million years ago and ended about 135 million years ago.

Laterite. A type of red soil that develops under wet, tropical conditions. Most laterite soils are very deep and also very infertile.

Lava. Magma that has erupted onto the earth's surface through a volcano.

Magma. A molten rock.

Marine Rocks. Rocks that formed in seawater.

Mesozoic. The era of geologic time comprising the Triassic, Jurassic and Cretaceous periods. Mesozoic time began about 225 million years ago and ended about 70 million years ago.

Metamorphism. The process of recrystallizing rocks under conditions of high temperature and pressure converting them to new kinds of rocks.

Metamorphic Rocks. Metamorphic rocks have been recrystallized under conditionsof high temperature and pressure. Most of them are coarse-grained and have a streaky appearance.Marble, schist and gneiss are good examples.

Mica. A family of common minerals which may be either black or colorless but are always flaky. Especially abundant in granites and similar rocks.

Miocene. The geologic period that lasted from about 25 to about 11 million years ago.

Mudstone. A sedimentary rock that started out as mud.

Obsidian. A glassy variety of rhyolite. Most obsidian is either black or brown and it usually forms large lava flows.

Oligocene. The geologic period that started about 40 million years ago and ended about 14 million years ago.

Olivine. A pale green mineral which occurs in small crystals scattered through black igneous rocks. Peridotite always contains olivine and so do some varieties of basalt and gabbro.

Peridotite. A heavy, black rock that forms most of the earth's interior. It is composedprincipally of black pyroxene and green olivine.

Perlite. A glassy form of rhyolite that contains some water. Most perlite is rather greenish but it comes in other colors. It puffs up like popcorn upon roasting to make lightweight chunks useful as a soil additive and in making special purpose concrete.

Plagioclase. A variety of feldspar which contains sodium and potassium.

Plate. One of the rigid slabs that make the outer crust of the earth. Plates are about 60 miles thick and most of them cover areas of many hundreds of thousands of square miles.

Pleistocene. The most recent 3 million years of geologic time during which the great ice ages occurred.

Pliocene. The time interval between 11 and 3 million years ago.

Pumice. A frothy variety of rhyolite glass. It is usually almost white and as light as wood.

Quartz. The commonest of all minerals. It comes in a variety of colors and disguises but usually occurs in clear, glassy grains. Quartz is the mineral form of silica.

Radiocarbon Dating. A method for determining the age of specimens of organic material by analysing their content of carbon-14 which is weakly radioactive. The method only works on objects less than about 40,000 years old so geologists rarely use it.

Rhyolite. A light-colored volcanic rock which usually comes in some pastel shade of gray, pink or yellow. Many volcanoes contain masses of solid rhyolite but elsewhere the rock generally occurs as deposits of pale volcanic ash.

Sandstone. A common sedimentary rock that was originally sand.

Serpentinite. A dark, greenish rock that is usually fairly soft and rather greasy looking. Many specimens feel soapy because they contain some talc. Serpentinite forms by the reaction of peridotite with water. It forms an important part of the oceanic crust.

Shield. A type of broad volcano with a very low profile composed mostly of basalt flows.

Silica. Silicon dioxide; when it occurs as a mineral, it is called quartz.

Sill. A thin sheet of igneous rock sandwiched between layers of sedimentary rock. They form when molten magma squirts between the sedimentary layers.

Tertiary. The period between the end of Cretaceous and the end of Pliocene time. The Tertiary period began about 70 million years ago and ended about 3 million years ago.

Triassic. The period of geologic time that began about 225 million years ago and ended about 180 million years ago.

Weathering. The complex of processes that combine to decompose solid rock into soil.

Zeolites. A family of minerals that most often occur as light-colored fillings in the gas bubbles of old lava flows. Most zeolites are light-colored and a few of them are rather attractive.

selected references

Baldwin, E.M., 1976, Geology of Oregon: Oregon University Bookstores, Eugene, Oregon, 290 p.

Beauleu, J.D., 1971, Geologic formations of western Oregon: Oregon Dept. Geology and Mineral Industries Bull. 70.

Beaulieu, J.D., 1974, Geologic formations of eastern Oregon (east of longitude 121 30"); Oregon Dept. Geology and Mineral Industries, 80 p.

Dole, H.M., editor, 1968, Andesite Conference Guidebook: Oregon Dept. Geology and Mineral Industries Bull. 62, 107 p. (articles on McKenzie Pass area, Crater Lake, Newberry Caldera, and Mount Hood).

Gilmour, E.H. and Stradling, D., editors, 1970, Procedings of the second Columbia River Basalt symposium, Eastern Washington State College Press, Cheney, Washington, 333 p.

McKee, B., 1972, Cascadia, the geologic evolution of the Pacific Northwest: McGraw-Hill Book Co., p. 139-254.

Ore Bin, Published monthly by Oregon Dept. Mineral Industries, 1069 State Office Bldg., Portland, Ore.

Peck, D.L., 1961, Geologic map of Oregon west of the 121st meridian; U.S. Geological Survey, Misc. Geol. Invest. Map I-325. 1:500,000.

Peck, D.L., 1964, Geology of the central and northern parts of the western Cascade Range in Oregon: U.S. Geological Survey Professional Paper 449, 56 p.

Peterson, N.V. and E.A. Groh, editor, 1965, Lunar geological field conference guide book: Oregon Dept. Geol. and Mineral Ind. Bull. 57.

Vallier, T.L., H.C. Brooks, and T.P. Thayer, 1977, Paleozoic rocks of eastern Oregon and western Idaho: in Paleozoic Paleogeography of the western United States, J.H. Stewart, C.H. Stevens, and A.E. Fritsche, editors, Soc. Econ. Paleont. and Mineral., Pacific Sec., Los Angeles, p. 455-466.

Walker, G.W., 1977, Geologic Map of Oregon east of the 121st meridian: U.S. Geol. Survey Misc. Geol. Inv. Map I-902. 1:500,000.

Williams, H., 1942, The geology of Crater Lake National Park, Oregon, with a reconnaissance of the Cascade Range southward to Mount Shasta: Carnegie Inst. of Washington Publication 540.

Index

volcanoes, 5, 6, 7, 8, 9, 22, 28, 38, 88,
    101, 109, 110, 111, 112, 114, 116,
    117, 118, 120, 121, 129, 142, 155

Wagontire, 240
Walla Walla, 110, 184
Wallowa, 253
Wallowa Lake, 253
Wallowa Mountains, 4, 6, 7, 10, 14,
    19, 20, 22, 25, 50, 53, 55, 161, 162,
    171, 248, 250, 253
Wallowa River, 250, 253
Warm Springs, 145
Warner fault, 241
Warner Range, 241
Wasco, 213
Washington, 15, 59, 82, 98, 117, 120,
    122, 161, 177
West Linn, 64, 101
Wilderville, 38

Willamette meteorite, 64
Willamette Pass, 149
Willamette River, 61, 73, 99
Willamette River bridge, 66
Willamette Valley, 15. 54, 58, 61, 62,
    63, 64, 66, 67, 68, 71, 73, 88, 99,
    101, 109, 127, 133, 140, 168, 176
Willamina, 99
Willow Creek, 208
Willowdale, 215, 216
Winston, 27, 108
Wizard Island, 157

Yachats, 89, 90
Yakima basalt, 168
Yamhill River, 99
Yapoah Cone, 138, 139
Yapoah flow, 139
Yaquina Head, 88

We encourage you to patronize your local bookstore. Most stores will order any title they do not stock. You may also order directly from Mountain Press, using the order form provided below or by calling our toll-free, 24-hour number and using your VISA, MasterCard, Discover or American Express.

Some geology titles of interest:

Title	Price
____ROADSIDE GEOLOGY OF ALASKA	18.00
____ROADSIDE GEOLOGY OF ARIZONA	18.00
____ROADSIDE GEOLOGY OF COLORADO	18.00
____ROADSIDE GEOLOGY OF HAWAII	20.00
____ROADSIDE GEOLOGY OF IDAHO	20.00
____ROADSIDE GEOLOGY OF INDIANA	18.00
____ROADSIDE GEOLOGY OF LOUISIANA	15.00
____ROADSIDE GEOLOGY OF MAINE	18.00
____ROADSIDE GEOLOGY OF MASSACHUSETTS	20.00
____ROADSIDE GEOLOGY OF MONTANA	20.00
____ROADSIDE GEOLOGY OF NEW MEXICO	16.00
____ROADSIDE GEOLOGY OF NEW YORK	20.00
____ROADSIDE GEOLOGY OF NORTHERN and CENTRAL CALIFORNIA	20.00
____ROADSIDE GEOLOGY OF OREGON	16.00
____ROADSIDE GEOLOGY OF PENNSYLVANIA	20.00
____ROADSIDE GEOLOGY OF SOUTH DAKOTA	20.00
____ROADSIDE GEOLOGY OF TEXAS	20.00
____ROADSIDE GEOLOGY OF UTAH	18.00
____ROADSIDE GEOLOGY OF VERMONT & NEW HAMPSHIRE	14.00
____ROADSIDE GEOLOGY OF VIRGINIA	16.00
____ROADSIDE GEOLOGY OF WASHINGTON	18.00
____ROADSIDE GEOLOGY OF WYOMING	18.00
____ROADSIDE GEOLOGY OF THE YELLOWSTONE COUNTRY	12.00
____GLACIAL LAKE MISSOULA AND ITS HUMONGOUS FLOODS	15.00
____AGENTS OF CHAOS	14.00
____COLORADO ROCKHOUNDING	20.00
____NEW MEXICO ROCKHOUNDING	20.00
____FIRE MOUNTAINS OF THE WEST	18.00
____GEOLOGY UNDERFOOT IN CENTRAL NEVADA	16.00
____GEOLOGY UNDERFOOT IN DEATH VALLEY AND OWENS VALLEY	16.00
____GEOLOGY UNDERFOOT IN ILLINOIS	15.00
____GEOLOGY UNDERFOOT IN SOUTHERN CALIFORNIA	14.00
____NORTHWEST EXPOSURES	24.00

Please include $3.00 per order to cover postage and handling.

Send the books marked above. I enclose $ _____

Name _____

Address _____

City/State/Zip _____

☐ Payment enclosed (check or money order in U.S. funds)

Bill my: ☐ VISA ☐ MasterCard ☐ Discover ☐ American Express

Card No. _____ Expiration Date:_____

Signature _____

MOUNTAIN PRESS PUBLISHING COMPANY
P.O. Box 2399 • Missoula, MT 59806 • Order Toll-Free 1-800-234-5308
E-mail: mtnpress@montana.com • Web: www.mountainpresspublish.com